Thomas Jefferson Twining

Genealogy of the Twining Family

ISBN/EAN: 9783337258702

Printed in Europe, USA, Canada, Australia, Japan

Cover: Foto ©berggeist007 / pixelio.de

More available books at **www.hansebooks.com**

Thomas Jefferson Twining

Genealogy of the Twining Family

GENEALOGY

OF THE

TWINING FAMILY

DESCENDANTS OF

WILLIAM TWINING, SR.

WHO CAME FROM WALES, OR ENGLAND, AND DIED AT
EASTHAM, MASSACHUSETTS, 1659.

WITH INFORMATION OF OTHER TWININGS, IN
GREAT BRITAIN AND AMERICA.

BY THOS. J. TWINING,
OF SIDNEY, INDIANA.

CHICAGO:
PUBLISHED FOR THE AUTHOR,
1890.

A NOBLE NAME.

I hold as reverend theme for rhyme
The name that glorifies its time;
A goodly heritage that will,
Through fresh inheritors, distill
Desire to widen wisdom's path,
Virtue, so given, to him who hath!

* * * * * *

A lineage old, of lustre new,
Moss-grown, yet green with latter dew—
This is the glory I would sing,
Until our children's children bring,
To match the name they won at birth,
A name of very present worth.

MARY B. DODGE.

There was a morning when I longed for fame,
There was a noontide when I passed it by,
There is an evening when I think not shame
Its substance and its being to deny;
For if men bear in mind great deeds, the name
Of him that wrought them shall they leave to die;
Thy duty and thy happiness were one.
Work is heaven's hest; its fame is sublunar,
Thy fame thou dost not need, the work is done.

JEAN INGELOW.

INTRODUCTION.

AN earnest desire to know something of his ancestors, their nativity, branching and history, led the writer, while yet within the borders of youth, to begin a course of inquiry thereto which has culminated in the present volume.

The object at first, was simply for personal gratification, the result of a reverence for the venerable forefathers who had lived, had their being; passing down the stream of time, to that limit from whence no voyager returns.

As the researches brought to light much that was deemed interesting and worthy of preserving, it was finally concluded to group, as near as possible, the descendants of the progenator who heads our records, into one family and give to the press for the benefit of whom it might concern.

The compiler felt unfitted for a work which has usually fallen to men of letters, in addition, his remoteness from the land of his fore-bears, the great libraries and sources of genealogical information, and having pursued the subject entirely within the bounds of his native state, through the medium of the pen alone, made the work in no small degree tedious and a little hopeless of final success.

Besides, as no other person, to his knowledge, was making an effort to trace the name in all its divergencies, it seemed incumbent upon him to search out the hidden and obscure records of generations dead and living. This volume is therefore sent out upon its mission, with the prayer that whoever reads it may be benefited. The desire is that its pages may awaken a deeper interest in the subject of Ancestry, a subject that too many regard lightly or know very little about. The beast of the field may have no conception of its own descent, but man, who rules over the animal creation, and of whose pedigree he manifests a concern not unreasonable, should certainly have a deeper interest in questions relating to his own and his children's lineage and the many important facts derived therefrom. If therefore a perusal of the following records shall be the means of lifting to higher and nobler deeds; if it assist in any measure to a humble and wisely spent life, and the acquisition of the moral and Christian principles that stand out so clearly in the lives of many who have passed from the stage of action to unmarked graves, the author will feel that his efforts have not been in vain. The amount of time and labor required to produce this little volume, will scarcely receive a just estimation by those not having engaged in a similar undertaking. Records of various kinds—church, state, county, historical, biographical and genealogical must be called into support; thousands of inquiries sent to every nook and corner where promise of information was apparent, and even the "cities of the dead" have been importuned for light. These all have required close examination and sifting, that the essential facts only be retained. Errors, the result of misinformation, or contradictory statements, and typographical, doubtless occur, however, the utmost diligence has been exercised to glean all the important facts relating to the numerous lines, and gladly would more have been added had the correspondents from whom we have largely drawn, seen proper to offer it. We would not weary the reader with a detailed account of years of wearisome labor, intervals of interruption, when hope for success was almost abandoned, and then recurring periods of delightful progress.

Thanks are due to all those who have given aid, and obligations are hereby acknowledged to Elizabeth H. Atkinson, (deceased), of Wrightstown, Pa., Mary B. Twining, of New Boston, Mass., Prof. E. H. Twining, of New York City, and Josiah Paine, Esq., of Harwich, Mass., without whose assistance there would have been seeming failure.

We have been content to confine our investigations almost exclusively to the first American Ancestor and his numerous descent, widely scattered over the northern portion of the United States, as far west as the Pacific ocean. It would be a pleasure, however, to see some one take up the subject, find the transatlantic connection, and trace the name in unbroken line on down to its earliest history.

THE NAME.

Regarding the family of TWINING, its origin, and significance of name, nothing is definitely known. Whether of Celtic or Anglo-Saxon beginning is not absolutely clear, although the burden of evidence thus far points to an Anglo-Saxon rather than a purely Welsh or Celtic origin. Traditions of the American family, with few exceptions, point to Wales as the place from which our ancestor came. This at the best would not prove a prior descent, and therefore leaves the question open to further enquiry and examination.

Having requested our friend, Prof. E. H. Twining, to write upon the subject, we are pleased to quote from him the following:—" The name of TWINING, like others ending in *ing*, is generally understood to be of Anglo-Saxon origin, and, as Richard Grant White says, 'of venerable antiquity.' Its significance is uncertain. One widely received view makes it a local appellation, derived from Saxon words meaning ' two meadows;' and regards the family as taking its designation from a locality in Gloucestershire, England, where, on the Avon, a few miles from Tewksbury, is a village bearing the name, and a ferry known as 'Twining's Fleet.' This is not, however, the view of Schele de Vere, nor of Isaac Taylor ('Words and Places,') both of whom, while not citing this name specially, state in general, that '—ing' was the usual Anglo-Saxon patronymic;' so that Waring, Harling and Twining rather gave names to places than received names from them. The word 'Ing' or meadow, according to the latter authority, is a local prefix only, as in *Ingham*. Of the two theories, in the absence of historic evidence, the weight of probability seems to lie on the side of the patronymic rather than the local. The Anglo-Saxon Chronicle (A. D. 547), gives:

'Ida waes Eopping,
Eoppa waes Esing,
Esa waes Inguing,
Ingui, Angenwiting,'

that is, 'Ida was Eoppa's son, etc.'"

"The term 'patronymic' is, in its strict sense, too narrow to express the force of 'ing,' which, like the Hebrew 'ben,' goes out to a wider range of relations; as in 'Viking,' *vik* (creek), *Haunter*, and 'cyning' (King), *the man of the Kin or Clan*.

"One member of the family, facetiously inclined, remembers that '*Twyn*' is Anglo-Saxon for *doubt*, and suggests a descent from the skeptical Thomas; another more seriously, finds his name among the mountains of Wales, in a word meaning *bush*. There are certainly Twinings in Wales, but translations of that kind are too infrequent to give countenance to even so plausible a supposition, without the support of direct evidence.

"In the matter of Christian names, *William* and *Thomas* appear everywhere, the latter often in combination with *John*. It seems that *William* and *John* were favorite names in England from the time of the Norman Conquest, and for obvious reasons. With the Puritan reaction the former fell into disrepute, as savoring of Paganism, and became infrequent among the reformers, and presumably in the direct ratio of their zeal. The spirit that could rejoice in the appellation of "Praise-God," or "Stand-fast-in-the-faith," was not likely to be led by any motives of sentiment to tolerate reminders of Gothic heathenism or Norman oppression. *John*, having good Scripture warrant, held its place, while *William* fell away, and only after the Restoration recovered its standing.

"The New England family, however, seems to have been, with the rest of the name, free from unreasonable prejudice in this regard, and retained the name, perhaps from a sort of kith-and-kin conservatism, perhaps from a dim sense of rhythmical fitness, along with *Thomas, Stephen, Elijah* and others drawn from the familiar Biblical readings.

So far as is known at present, there are in Great Britain three points at which the name of TWINING appears to have had early, it may be independent, foothold, in Gloucestershire, as already mentioned, in Wales, and in the vicinity of St. Mary's Isle, Kirkcudbrightshire, Scotland.

"In Gloucestershire, John Twining appears as Abbot of Winchcombe, about the middle of the fifteenth century; and in the Scottish locality the name is said to be found on tomb stones of at least as early a date. Families of the name exist at present in Scotland and Wales, in London, and elsewhere in Middlesex, and perhaps at other points in England. On this side of the Atlantic they are met with in Nova Scotia,* coming originally from Wales, and throughout the United States.

* The name also occurs in Newfoundland and the Canadas, as well as Nova Scotia, where it has had an existence for some hundred years, the ancestor being a native of Wales. These different families are not very numerous, and whether they all centre in the Welsh family is not known. They are characterized as refined, well-informed, respectable and well-to-do people.

Richard Twining, of London, (editor of Papers of the Twining Family, by Rev. Thomas Twining, at one time Rector of St. Mary's, Colchester, England), holds to the theory that the name had its origin at the junction of the Severn and Avon, from whence it is claimed members of the family emigrated to other countries.

"These last, with the exception of two or three additions, the earliest of which dates its arrival in this country not more than thirty years back, seem to be all descended from one ancestor.

"**Mr. William Twining**, who first appears in public records, as a freeholder at Yarmouth, on Cape Cod, Massachusetts, in the year 1643, though there is reason to believe that he came over at a somewhat earlier date.

"There is no light upon the particular circumstances which led our ancestor to leave his home, and no direct information as to the part of Great Britain from which he came; but there can be little question that the influences which moved him were the same as, in one or another form, had been operating throughout under the rule of the Stuarts, and even under the last of the Tudors, to detach Englishmen from their native soil. These were, briefly, church intolerance, court corruptions and royal exactions. That he did not come earlier may have been due to a survival of loyalty, or to conditions of place or occupation which brought the pressure of these influences less directly upon him than upon many others, or to a temperament which, hopeful of better things, held him at home until it seemed that the success of neither of striving parties promised any good. The tyranny which the Presbyterians forced upon Cromwell was quite as insufferable as the Erastianism of Laud; and while the Puritan element in general was drawn by its moral sense into co-operation with the Presbyterians, those who, like the Pilgrims before them, more broadly asked for freedom of faith and practice, seeing that presbyter was, as John Milton phrased it, only 'priest writ large,' found themselves between two fires, and 'turned to the New World to redress the balance of the Old.'

"Of the twenty thousand or more who emigrated between the years 1629 and 1640, the time of arrival of only a relatively small number can be ascertained from the passenger lists of the vessels on which they sailed. If William Twining came after the proclamation prohibiting emigration without license (May 1, 1638), and prior to 1640, when emigration had practically ceased, it is not difficult to see why his name might not appear on the registers. In the first place, although ships left England for America almost daily, Hotten's lists give names of passengers of but one ship in 1638 and 1639. It is certain registers were kept, but equally certain that nearly all of them have been lost.

"Further, these registers contained only the names of those who left England *legally*, i. e., under license according to the proclamation, and doubtless thousands left secretly to avoid the oath of Allegiance and Supremacy and the payment of a subsidy to the crown, as well as to escape the annoyances and disabilities which attended those who were disaffected to the Church establishment.

"If he came over after 1640, in November of which year the Long Parliament assembled, he could perhaps have come without hindrance, and without official registry. Why he should have emigrated at that time is a matter of conjecture. A possible answer to the question has been already intimated, but so far nothing 'of record' has been discovered to settle it definitely. Possibly some evidence may exist in England to throw light on the true cause of his migration and that of others who came under the same circumstances. Perhaps, too, the fact of his joining the Yarmouth Settlement (which after an attempt in 1638 was incorporated in 1639) may have been a consequence of the presence there of some friends from the Old World, and this might give a clue to the port from which he sailed. It has been assumed that he came from an English locality. The reason for this assumption is that, so far as is known and as might be expected, the greater number of Colonists came from those parts from which the seaports were most accessible, and which at the time were most easily reached, by the oppressions and exactions which stimulated the movement. Further, it appears that the immediate descendants of William Twining were decided Congregationalists, and one branch of the family affiliated with the Society of Friends. It may be inferred that he was himself of the independent Puritans, and so out of touch with the royalists faction on one side and the Covenant on the other. As has been intimated above, this may have been a reason for his emigrating at a time when no general movement was taking place. It is known that there were Twinings who adhered to the cause of King Charles, and family differences may have had something to do with his expatriation. These are, however, but matters of conjecture, and may serve to indicate lines of future exploration. What may be safely held for fact is that the ancestor of the American family was Anglo-Saxon and Puritan. The race-type remains in the family to-day—the sanguine temperament, the fair skin, the blue eyes, (often with the intent look that led Tacitus to call them truces), which the Romans saw eighteen centuries ago, and the same love of honest dealing and fair play. That he was Puritan hardly needs proof. It would require strong evidence to show that he could have been other, when the whole middle and professional class in England was Puritan.

"The opening of the Bible by Elizabeth had set up in the English people a new standard of life and conduct, and stimulated a reaction against the caprice and passion and careless geniality of the age. The aim of the Puritan 'was to attain self-command, to be master of himself, of his thought and speech and acts;' and this not as a social art, but under stress of devotion to a Supreme will. This temper was likely to lead to a narrow and ascetic habit, lacking grace and color of life; but it had the definiteness and efficiency that belong to intensity of conviction and singleness of purpose. Our ancestor has left no words or history to show these traits of character; but they are claimed by the traditions of the family; they are held as an inheritance, and their imitation as an obligation; for 'the growing good of the world is partly dependent on unhistoric acts.'"

NOTES.

—Bucks County, Pa., has been the banner home for Twinings for near 200 years; for a like period, Barnstable County, Mass., was the family abode, but has been without a representative of the name for many years.

—Berkshire and Hampden counties, Mass., have a few reminders of the once fruitful family. Tolland, at one time noted for its Twinings, has passed the name almost to the point of obscurity.

—Broome County, N. Y., retains her reputation for numbers. In Hancock County, Ohio, the name was legion, but removals have depleted the ranks. The name has had a prominence at New Haven, Ct., since 1795, and Granville, O., since 1815.

—Philadelphia and Scranton, Pa., are chief Twining centres. Pennsylvania, New York, Ohio and Connecticut, in the order named, contain a majority of the living generations. Massachusetts, Iowa, Illinois, Michigan, Wisconsin and Kansas are about equally balanced in families. In Indiana, Vermont and New Hampshire the name is nearly extinct.

—It is observable that the New England Twinings are pretty much "scattered," while the Pennsylvania families seem to preserve a more compact form; this may in measure be owing to the latters greater numbers, and the communistic trend of their Quakerism.

—Occasionally the two great lines intersect, though it does not appear that they have ever intermarried. New York and Ohio are States where the two branches are brought in closest relationship.

—Ministers, physicians, lawyers, civil engineers, and school teachers are adequately represented; farmers predominate, not many wealthy, but usually the Twinings are a well to do people, observers of law and the rights of their fellow man. No statesmen or men of renown.

—The religious faculty is prominent, inclined to the liberal, progressive; in politics largely Republican; in temperament it predominates in the nervous-sanguine. When the fire is wont to fly, "dog nor devil" is seldom taken into account; a calm follows quickly.

—In their love for country the Twinings have abundantly shown their patriotism in the defense of home and on the field of conflict; not a few of them have died in battle and hospital. One served in the war between France and the Provinces, 1758. One is known to have served throughout the Revolution, one or more in the 1812 war, and one was so unpatriotic as to serve in the Southern Confederacy.

—To the fifth generation the families were confined to the narrow limits of Cape Cod and Bucks County, Pa. From thence, about 1780, they began extending their lines, the Cape family to Tolland, and the Pennsylvania family to Philadelphia, New Jersey and York State. About 1793 a leading branch of the family "down on the Cape" settled at Frankfort, now Winterport, Maine.

—In a long and extensive research there have not been found in the United States a half-dozen Twining families who were not identified in the American progenitor. The English name occurs at Chicago and Denver; the Nova Scotia (Welch) Staten Island, and the Scotch at New Haven, Ct. A family bearing the name also resides at Otis, Mass.

—The branches which it has been impossible to explore and which the future genealogist may be able to extend, are: *Nathaniel*, Jr., son of **6** Nathaniel; *Samuel*, grandson of **20** Barnabas; *Christopher*, son of **25** Daniel; and *Eleazer*, son of **81** Barnabas. The latter two, it is quite evident have descendants.

—A question of heirship not infrequently arises over vast and unclaimed estates in England. One of the eighth generation tells of hearing her grandfather (b. in Pennsylvania) say he had a rich old bachelor brother in England (?), and she wonders if he is not one of the two, Richard Twining and Wm. H. Twining, reputed millionaires, whose vast estates are said to be awaiting claimants. This is a fair specimen of many traditions afloat, and unless members of the family desire to throw away their money it is not advisable to put in claims upon foundations so untenable.

—To clear up a few of the traditional stories in vogue among descendants, some of which are that "three brothers" came over, and a few that there were two, no evidence has been presented indicating that Wm. Twining, Sr., had any connection other than his own family. Nor can it be found that any other person of the name Twining ever set foot on Colonial soil. Unsupported and diverse statements are frequently thrown in the way of the earnest enquirer, but are not the "stubborn things" which he is wont to feed upon, and must be cast aside for that which is true.

—About 1845 the Barnstable county, Mass., Records, were all destroyed by fire, and many wills and documents of the early families were thereby lost, and may account for our Progenitor having no records extant, which might shed light on several uncertain questions of lineage, marriage, issue and worldly estate of the latter, of which no facts are given.' The Congregational Church records of Orleans exist, according to Barber's Hist. Coll., from 1648 to 1797; according to Josiah Paine, none extant before 1773. The Eastham Church Records all lost.

—The early Plymouth Records show some variation in spelling the name. We find it Twining (right), *Twineing*, *Twiney*, and *Twyneing*. The index of same records treat them all as one name.

—Rev. Kinsley Twining, D.D., regards the name of Welch or Celtic origin, and therefore hard to catch and identify. He says: "Thomas Twining, of London, evidently did not consider the name worth bragging about, though they were considerably enlightened to have a coat of arms. He, however, changed the crest and started anew with one contrived out of an entirely false rendering of the name, which evidently he supposed to be some derivation of *twine*, and accordingly, as his grandson or greatgrandson, the late distinguished banker, explained to me, provided himself with a crest showing an arm or something of that kind with a serpent twisted about it. All this proves that he knew nothing of the real significance of the name."

THOS. J. TWINING.

EXPLANATIONS AND CONTRACTIONS.

The plan of the book is not original, but has been adopted with a few modifications, as being the least confusing to the ordinary reader, and avoiding repetition of female branches, which are recorded for but *one generation*. It is not customary to trace the female lines beyond the children, for reasons that the original name becomes suspended, and also that the labor required to follow these thousands of descendants would seem boundless.

The heads of principal families and branches, are numbered in succession from 1 TO 175; the name of the person numbered being preceded by *heavy figures*. Names which are not preceded by *heavy figures* are completed in one notice, and show that said name has no living *male* issue, except in case where said issue is lost or not in record. The small figures number the children of a family. For uniformity, dates of births, deaths and marriages, are given in the usual form of the times. In abbreviating, *b.* stands for born; *d.* for died or deceased; *m.* for married; *unm.* for unmarried; *inf.* for infant; *dau.* for daughter; *ch.* for child or children; *res.* for residence; *Tp.* for township, etc. Where the name of a place has been given once its initial will be understood thereafter, as *E.* for Eastham; *O.* for Orleans; *T.* for Tolland; *N.* for Newtown and *W.* for Wrightstown, etc.

GENEALOGY.

Genealogy or " The history of the succession of families," is a very old word, being employed in sacred and profane history.

Of the term,—its usage and history, we quote from *Zell's Encyclopedia and Dictionary*.

" Genealogy, jen-e-ålo-je, n, Fr. génèalogic; Gr. *geneologia*, from genea, a race, and *logos*, a discourse, an account or enumeration of the ancestors or relations of a particular person or family. No nation was more careful to trace and preserve its genealogies than the children of Israel. Their sacred writings contain genealogies which extend through a period of more than 3,500 years, from the creation of Adam to the captivity of Judah, and even after that time. Josephus informs us that he traced his own descent from the tribe of Levi by means of public registers, and that, however dispersed and depressed his nation were, they never neglected to have exact genealogical tables prepared from authentic documents which were kept at Jerusalem. Since, however, their destruction as a nation by the Romans, all their tables of descent seem to have been lost; and even the Levites, who are still distinguished from the rest of the people by the exercise of special honorary religious functions, are known as such only by being acknowledged as descendants of parents who exercised the same.

The inequalities of rank and right which prevailed during the Middle Ages, made genealogical inquiries highly important, and it was then that researches of this kind assumed the form of a science, which became closely connected with heraldry, q. v. Very little critical care, however, was usually employed in such cases, the chief object being to trace the origin of families into the remotest antiquity.

Attempts to carry this to an absurd length are frequently manifested in the earlier genealogical works. Critical genealogical studies were not begun before the Seventeenth Century.

Genealogical accounts are not only interesting to persons who feel a more or less natural curiosity about their ancestors, but are also useful to the historian, as elucidating the often complicated relations of dynasties, families, claims and controversies of successions, etc. They are also important in legal cases concerning claims of inheritance, and, indeed, are indispensable in States in which the enjoyment of certain rights is made to depend upon lineage or descent.

A genealogy, or lineage, is frequently represented in the form of a tree, *arbor consanguinitalis*, giving a distinct view of the various branches of the family, and the degrees of descent from the common progenitor, who is generally present in the root or stem.

Genealogical tables are either descending or ascending. The former are chiefly used in historical records, presenting the descendants of a certain person in the order of procreation; the latter in documents of nobility, serving to show the claims of any man or family to the titles of paternal and maternal ancestors.

Sometimes both forms are used together. Persons descended one from another successively, form a direct line; those descended from a common progenitor, but not one from another, a collateral line; the collateral line embraces the *agnates*, or the kindred on the father's side, and the *cognates*, or kindred on the mother's side."

In a work entitled, " Two Sticks," or, " The Lost Tribes of Israel," the author attempts to prove the Anglo-Saxon race to be the ten lost tribes. Bearing on the subject of genealogy, he says: " What means this effort, both in England and in America to trace up genealogies? Why so much anxiety to know the lineage? Is it because there is a prospective inheritance in the Holy Land? Are Israelites desirous of knowing what share of the new country is theirs?"

* * * * * * * * * *

" Men may attribute this hunting of ancestors to an ' idle curiosity,' a ' fancy of the brain,' still there may be a providence behind it. Thousands have performed work which, at the time, was considered as having no significance from a divine standpoint; but what have seemed mere fancy to the multitude turned out to be the hand of God in shaping the destiny of men and nations.

* * * * * * * * * *

" How often has that which has been considered as mere happening turned out to be ' the finger of God?' Now, while it may *seem* to be improbable that this seeking of genealogies has any relation to Israel's gathering, it might be well to place it in the catalogue of portentious events that are transpiring in perfect harmony with prophecy, divine utterances from the glory and presence of God. When events and predictions fit so happily, it would be rudeness to tear them asunder. The children of Israel were very careful to preserve their genealogies."

HEADS OF FAMILIES.

WHICH ARE NUMBERED IN THIS BOOK.

1 William, first American Anc'r.
2 William, of **1** William.
3 Stephen, of **2** William.
4 William, of **2** William.
5 Stephen, of **3** Stephen.
6 Nathaniel, of **3** Stephen.
7 John, of **3** Stephen.
8 William, of **4** William.
9 Barnabas, of **4** William.
10 Stephen, of **5** Stephen.
11 Samuel, of **6** Nathaniel.
12 Benjamin, of **6** Nathaniel.
13 John, of **7** John.
14 Eleazer, of **7** John.
15 Jacob, of **7** John.
16 Stephen, of **7** John.
17 Thomas, of **8** William.
18 Elijah, of **8** William.
19 Jonathan, of **9** Barnabas.
20 Barnabas, of **9** Barnabas.
21 Prince, of **9** Barnabas.
22 Stephen, of **10** Stephen.
23 Thomas, of **11** Samuel.
24 John, of **11** Samuel.
25 Daniel, of **12** Benjamin.
26 Joseph, of **13** John.
27 Silas, of **14** Eleazer.
28 David, of **14** Eleazer.
29 John, of **15** Jacob.
30 Jacob, of **15** Jacob.
31 David, of **15** Jacob.
32 Jacob, of **16** Stephen.
33 Stephen, of **17** Thomas.
34 William, of **17** Thomas.
35 William, of **18** Elijah.
36 Eleazer, of **18** Elijah.
37 Judah, of **18** Elijah.
38 Lewis, of **18** Elijah.
39 Nathan, of **19** Jonathan.
40 Barnabas, of **19** Jonathan.
41 Abner, of **20** Barnabas.
42 Jonathan, of **21** Prince.
43 Prince, of **21** Prince.
44 Charles, of **22** Stephen.
45 John, of **23** Thomas.
46 Charles, of **23** Thomas.
47 Thomas, of **23** Thomas.
48 Thomas, of **24** John.
49 John, of **24** John.
50 Samuel, of **24** John.
51 Benjamin, of **24** John.
52 Mahlon, of **24** John.
53 Benjamin, of **25** Daniel.
54 Jacob, of **25** Daniel.
55 Jacob, of **26** Joseph.
56 John, of **26** Joseph.
57 Joseph, of **26** Joseph.
58 Watson, of **27** Silas.
59 Silas, of **27** Silas.
60 William, of **28** David.
61 Isaac, of **28** David.
62 Thomas, of **28** David.
63 Jacob, of **29** John.
64 Abbott C., of **29** John.
65 Isaac H., of **29** John.
66 Jessie B., of **30** Jacob.
67 Henry M., of **30** Jacob.
68 Cyrus B., of **30** Jacob.
69 Amos H., of **31** David.
70 George, of **31** David.
71 Croasdale, of **32** Jacob.
72 Stephen, of **32** Jacob.
73 Alexander C., of **33** Stephen.
74 William, of **33** Stephen.
75 Alfred C., of **34** William.
76 Alexander H., of **34** William.
77 William, of **35** William.
78 Elijah, of **35** William.
79 Hiram, of **35** William.

THE TWINING FAMILY.

80 Joseph, of **35** William.
81 Barnabas, of **36** Eleazer.
82 Philander F., of **37** Judah.
83 Merrick S , of **38** Lewis.
84 Edward W., of **38** Lewis.
85 Jonathan, of **39** Nathan.
86 Ebenezer, of **40** Barnabas.
87 Addison, of **41** Abner.
88 Harrison, of **41** Abner.
89 Jonathan, of **42** Jonathan.
90 Stephen B., of **44** Charles.
91 Edward W., of **44** Charles.
92 Aaron, of **45** John.
93 Nathan C., of **45** John.
94 Henry H., of **45** John.
95 Peter S., of **45** John.
96 Chapin, of **46** Charles.
97 Dewitt C., of **47** Thomas.
98 Lewis, of **47** Thomas.
99 Charles, of **48** Thomas.
100 James, of **49** John.
101 Thomas, of **49** John.
102 William, of **49** John.
103 John A., of **49** John.
104 Charles, of **49** John.
105 Philip, of **49** John.
106 Joseph N., of **50** Samuel.
107 Charles A., of **50** Samuel.
108 David M., of **51** Benjamin.
109 Joseph, of **52** Mahlon.
110 Chester P., of **52** Mahlon.
111 Frederick F., of **52** Mahlon.
112 William F., of **52** Mahlon.
113 George R., of **52** Mahlon.
114 Mahlon J., of **52** Mahlon.
115 Henry L., of **52** Mahlon.
116 Jessie, of **53** Benjaman.
117 Eli, of **53** Benjamin.
118 John, of **53** Benjamin,
119 William, of **53** Benjamin.
120 Horace G., of **53** Benjamin.
121 Ralph, of **53** Benjamin.
122 Samuel, of **54** Jacob.
123 Malichi, of **55** Jacob.
124 Joseph, of **55** Jacob.
125 John, of **55** Jacob.
126 James, of **55** Jacob.
127 Jacob, of **55** Jacob.
128 Ralph L., of **55** Jacob.
129 Silas, of **56** John.
130 Jonathan R., of **57** Joseph.
131 Hallowell S., of **58** Watson.
132 Elias B., of **58** Watson.
133 Silas, of **59** Silas.
134 Uriah R., of **60** William.
135 William W., of **60** William.
136 D. Hallowell, of **61** Isaac.
137 Horace B., of **61** Isaac.
138 B. Franklin, of **61** Isaac.
139 Charles H., of **62** Thomas.
140 John, of **63** Jacob.
141 Isaac H., of **63** Jacob.
142 John W., of **64** Abbott C.
143 Thomas C., of **64** Abbott C.
144 David R., of **65** Isaac H.
145 Howard L., of **67** Henry M.
146 Jonathan A., of **68** Cyrus B.
147 Wilmer A., of **68** Cyrus B.
148 William H., of **69** Amos H.
149 John, of **69** Amos H.
150 Edwin of **71** Croasdale.
151 Kinsley, of **73** Alexander C.
152 Edward H., of **74** William.
153 Charles O., of **74** William.
154 George A , of **75** Alfred C.
155 William F., of **76** Alexander.
156 John, of **77** William.
157 William F., of **77** William.
158 Alfred W., of **77** William.
159 Elphonzo, of **78** Elijah.
160 Joseph, of **78** Elijah.
161 Orlandon, of **78** Elijah.
162 Samuel M., of **78** Elijah.
163 Lucius, of **78** Elijah.
164 Samuel R., of **79** Hiram.
165 Nelson B , of **82** Philander F.
166 Homer P., of **82** Philander F.
167 Lewis S., of **83** Merrick S.
168 Edward W., of **83** Merrick S.
169 Henry L , of **83** Merrick S.
170 Nelson L., of **83** Merrick S.
171 Edward T., of **84** Edward W.
172 Lauriston, of **84** Edward W.
173 Jessie L., of **84** Edward W.
174 Hiram, of **85** Jonathan.
175 George F., of **89** Jonathan.

FIRST GENERATION.

1 William Twining, Sr., is the central figure around which cluster a multitude of living characters, comprising ten generations of descendants, measured by a distance of nearly two hundred and fifty years.

This important personage, as he first appears in New England history, on the shore of Cape Cod, awakens a deep interest in the minds of those who reverence antiquity and the hallowed ground of ancestral abode.

Of his nativity nothing is known to certainty. From whence he came, his paternity, life and traits of character, the vessel in which he took passage and its date of arrival in the New World, are questions which have been pondered over and the page of history examined that some ray of light might shed her pleasant beams and answer fully and satisfactorily. These questions may never be solved. The impenetrable darkness of unrecorded events may never declare his life prior to his voyage across the Atlantic. Indeed, his milestones seem to be hidden in the dead past, and the reader can but ponder as over other problems in the world's history which the intelligence of man can but feebly grasp in speculation.

We say that reliable information regarding our ancestor is wanting; family tradition, however, almost uniformly assert that he came from Wales. One that he came from Yorkshire, England, and one, an aged spinster living in the vicinity of Eastham, speaks of a "taint of French blood."

The utmost that has been gleaned from the various records and sources of information relating to historical and genealogical matters, as to time of his entry into the New World, is taken from James Savage, an authority on New England genealogy, who states that "Mr William Twining, Sr., was in Yarmouth 1643."

Other New England writers, who doubtless copy from Savage, make substantially the same statement.

The Yarmouth records imply that he was there at least two years earlier, or 1641, the date of his daughter Isabel's marriage.

It is questionable whether he first landed at Yarmouth. We are fain to believe that he first touched shore at Plymouth and was among those of her settlers who became dissatisfied with their location and sought new homes at various places along the Cape Cod Coast. That he was a man of more than ordinary character is shown by the title which prefixes his name in the early records, an appellation of honor which was rarely applied in those days, as shown by the History of Mass. Bay, which tells us that "the first settlers of these Colonies were very careful that no title or appellation be given where not due. Not more than half a dozen of the principal gentlemen of the Massachusetts Colony took the title of *Esq.;* and, in a list of one hundred freemen, not more than four or five were distinguished by a *Mr.*, although they were generally men of substance. Goodman and goodwife were the more common appellations."—*See Hutchinson's History.*

Additional to this, another historian says, referring to the changes at Plymouth and the standing of those who first came to Eastham: "The church at Plymouth regretted their departure, for they who went out from her were among the most respectable of all the inhabitants of Plymouth."

Tracing his future, chronologically, we find: In 1643 William Twining is included in a "list of those able to bear arms" at Yarmouth. The Y. records 1643-5 rank him among the militia, which consisted of fifty soldiers. In 1645, he was one of five soldiers sent out against the Naragansetts. (Ply. Col. Rec., Vol. II, page 91). The next heard of him is at *Eastham, where Savage says he removed soon after being at Y. 1643. As Governor Thomas Prence went to E. 1644, it is more than likely William followed soon after 1645. The first mention found of him at E. is taken from the original book of records (Vol. 2) of Eastham, wherein he is put down as constable, June 5, 1651; however the Ply. Records (Vol. 2, p. 167), say " elected Constable Eastham 5 June, 1651, William Twining, *Jr.*" It is not certain therefore that the *Sr.* was meant in the former, although he was doubtless intended in both entries.

<small>*Eastham, called Nauset prior to 1651, in its original form, is a township of Barnstable County, Mass., having the Atlantic Ocean on the East, 15 miles long and 2½ wide: present length 6 miles. It is almost a continuous plain; soil sandy, requiring much labor. Portions of it yield large crops. Fresh water ponds and several creeks exist in the township. Excepting a tract of oaks and pines about ½ mile wide, no wood remains in the Tp. Business is agricultural and maritime. Fishing has disappeared entirely. The public buildings are four school houses, a town hall and Methodist Meeting house. The old Cong. Church has become extinct, after an existance of more than two centuries. Settlement of town began 1644, with 49 souls, who soon built a meeting house 20 ft. sq. with thatched roof and forts in the side of building, for use in case of attack by the Indians. Around this house, near Town Cove, a burial place was laid out, and still enclosed, containing some grave stones.

Rev. Jno. Mayo was the first pastor 1646. Wolves were plenty; the bounty on four killed in town was paid 1655. In 1663 all horses were ordered branded with the letter E., and all persons who stood outside of M. house during public services to be set in the stocks. In 1667, every housekeeper was required to kill a certain number of blackbirds and crows. In 1684, 101 legal voters reported and 500 adult Indians within the parochial charge, 1718, Tp. divided into North and South Parish. In 1758, an inhabitant of the N. Precinct having avowed himself a Baptist, the parish voted to remit his tax for support of the Cong. Ch. In 1765 but four Indians in E. In 1810 a Methodist society was established, shortly before, every inhabitant in the Tp. belonged to the Cong. Ch. These selections from the " Annals of E.," by Dr. Frederick Freeman are made, as very many of the Twinings, who are scattered abroad look back to this " cradle of an honored race," as the place where they themselves " have drawn the principles and habits that have made them prosperous and honored wherever they have gone to seek their fortunes in the wide world." For two centuries this township was made a place of dwelling by some of the descendants of William Twining, but now only a few of the female des. remain.</small>

To the Orleans records is due the statement that William, m. Annie Doane* 1652, and that she d. Feb. 27, 1680.

In 1655, he had granted him, or his purchases confirmed, and under same date we have " propounded to take up their freedom * * * * * * William Twining," by which it is presumed the Senior is indicated.

His death occurred at E. April 15, 1659, and he was probably not more than 65 yrs. old, if we may approximate by contemporaneous and circumstantial evidence.

That he was a member of the Congregational Church seems conclusive, yet no facts at all can be given in proof.

No data is at hand regarding his first marriage and it is simply conjecture whether his wife died before he came to "our shores" —probably soon after. Authorities from whom meagre light is drawn on the subject, had no certain knowledge of a *second marriage*, although Savage does say, "it is not certain that she (Ann) was the mother of Wm. **2nd**, as she may have been 2d wife of Wm. **1st**."

There is no evidence of any children by this 2d marriage and nothing definitely known of the full fruits of *first* m. Certain it is he had:

1.—**2 William (2nd** or Jr.), whom it is asserted was probably b. in England and m. Elizabeth Deane.
2.—**Isabel** m. Francis Baker, June 17, 1651. Mr. B. came over in the Planter, in 1635, at the age of 24, from Gt. St Albans, Hertfordshire, Eng.; went from Ply. to †Yarmouth, about 1642, and with the Twinings to E., but returned to Y. where he d. before 1700; will made Mch. 4, 1693. The Court Rec. tell us he was a Cooper by trade and that 1648 he was made surveyor of highways for Y.,

† Annie was probably a sister of Dea. John Doane, b. 1590; came from Wales to Ply. 1621, one of the first founders of Eastham, and assistant of Gov. Prence of Ply. 1633. He d. 1685, the head of a numerous and respectable descent, whose genealogy is in preparation by Jno. A. Doane, of Atlanta, Ga.

1653, presented by G. J. "for retailing wine contrary to order of Court;" 1655, likewise, for abusing a servant. In 1657 took oath of fidelity; 1665, fined for breach of peace, and 1680 he sues Abraham Hedge for £12, for tar barrels.

* The early records of Yarmouth, to 1671, are lost.

ISSUE.

1. Nathaniel, b. Mch. 27, 1642. He and wid. d. 1691.
2. John, b. Yarmouth.
3. Samuel, b. May 1, 1648.
4. Daniel, b. Sept. 2, 1650.
5. William.
6. Thomas.
7. Eliza, m. John Chase.
8. Hannah, m. —— Pierce.

SECOND GENERATION.

2 William Twining, son of William Sr., was b. in early part of seventeenth century, Savage says, "probably in Eng.," and died in Newtown, Bucks Co., Pa., Nov. 4, 1703; m. Elizabeth Deane (dau. of *Stephen of Ply., whose wid. m. Josiah Cooke) at E. about 1653, possibly several years sooner. She d. in N. Pa , Dec. 28, 1708. The court records mention him first, June 3, 1652, when he was admitted and sworn; 1652 he was one of the Grand Jury or Enquest; 1667-8 on same, to examine as to a child found dead in the woods, etc.; June 5, 1671, sworn as one of the Grand Enquest.

He was a deacon at the Eastham Church, as early as 1677. Freeman nor Platt, in their histories of Cape Cod, do not mention this fact, however, the old clerk says, under date Feb. 26, 1677, "Deacon Twining and dea. Freeman," "to take and keep a clear account of Mr. Treat's maintenance given him by the town," "Dea. T. and dea. F. to take care about underpining the house that Mr. Treat now lives in, to see that it is sufficiently done, etc. Again he is alluded to as "deacon Twining" as late as 1681. From old records of E.:—

"May 5, 1693, Granted to Joseph Young, a

---

* "Stephen Deane of P., one of the first comers, in the Fortune, 1621, built the first corn mill in N. E. 1632; m. about 1627, Elizabeth, dau. of wid. Ring: had Eliz., Miriam and Susanna, and d. Sept. 1634. His wid. m. Josiah Cooke and d. 1687" "Andrew Ring of P. (whose wid. mother made her will there in 1633), came probably about 1629: was freeman 1646. His sister Elizabeth who m. Stephen Deane, was older than Andrew, whether the father came or died in England is not certain."

parcel of land at "Poche,† Capt. Sparrow and William Twining, Sr., to set it out." He was a proprietor of land at "Easton Harbor," and had interest in Drift Whales at the end of the Cape.

In 1695, he and his son 4 William were enumerated among the legal voters of E. Near the latter date, his religious views seem to have underwent a radical change. He has evidently became convinced of Friends' principles and now contemplates ‡removal to the newly settled Province of Penn., where the tenets of Quakerism were maintained in their purity, and freed from the intolerance of New England theology. The records testify that there were Quakers in Eastham, but it cannot be found that they held Monthly Meetings in the town. They doubtless belonged to the Sandwich Society, which was organized very early. Although this Society makes no mention of William's name upon its minutes, it is most probable he became, as also his wife and son Stephen and his family, identified with said Society prior to removal.

1695 is the year in which he bids adieu to the land of the Pilgrims. It marks an important epoch in the family history. Hitherto the name appears to have borne honor to the Cong. Church alone. Up to this date, a period of near six decades, the family was confined on the historical Cape Cod; but now the house is about equally divided, religiously and geographically, and henceforth it is Quaker and Cong. blood flowing in parallel lines from generation to generation.

† "Poche" or 'Pochet,' is the Indian name of that portion of old E., now known as East Orleans. It is a very large territory, almost entirely surrounded by salt water. It is now known to some extent as Pochet. It was a famous resort of the red man, and was very early settled by the whites, on account of the good soil. The old cemetery is just on the border of the locality. The Twinings settled on the Neck. To 'to set out' land—a term used by our early English settlers—meant *laying* out a parcel."

‡ From about 1670 to 1700 there was a large emigration from different parts of New England to East Jersey, and later to Penna., especially with Quakers, and some other "Heretics,"—on account of the persecution.

"Deacon Twining," now having donned the Quaker garb, speaks the Quaker tongue, becomes the unretaliative friend of the Indian, whom he seeks to elevate. He is in fact a believer and exponent of all that is comprehended in the teachings of Penn, Fox and Barclay. Upon those of his descendants from this new home in the wilderness of Penn., on the banks of the Delaware River, were stamped those newly acquired principles which time nor space have failed to efface.

Owing to inadvertancy, Freeman is led to say that two of William **2**d daughters were the children of WILLIAM **1**st and his wife ANNE, namely: ELIZA and ANNE.

This is correctly shown in the subjoined will.

"I, William Twining, of Newtown, in the county of Bucks, and the province of Pennsylvania, having taken into consideration the frailty of this temporal life, being in health of body & of sound and perfect mind and memory, praise be therefore given to Almighty God, do make & ordain this my present last will & testament, in manner and form following (that is to say):

First & principally I commend my soul into the hands of Almighty God, & my body I commit to the Earth, to be descently buried at the discretion of my executors hereafter named, and as touching the disposition of all such temporal estate, as it hath pleased Almighty God, to bestow upon me, I give & dispose thereof as followeth:

First, I will that my debts and funeral charges shall be paid and discharged.

Item. I give unto William, son of Stephen Twining (My Grandchild) ye sum of twenty Pounds, to be paid to him when he shall come to twenty-one years of age, if my wife be dead before he comes to that age, but if she be living I leave it with her, for her use so long as she liveth, and then to be paid at her decease to Said Wm. Twining.

Item. I give unto my son Stephen Twining, my daughter Johanna, my daughter Mahitable, and the two daughters of my daughter Anna Bills (to wit) Anna Bills & Elizabeth Bills, as ye residue of my estate which I have in Penna, after my wife's decease (that is) my mind is, that Anna Bills & Elizabeth Bills, have both but an equal share with my son Stephen & the rest of his sisters.

Item. I give unto my son William Twining (after my wifes decease) all the lands, or the residue that I have lying within the Township of Eastham, and County of Barnstable, in New England, so far as ye* Bound Brook, lying on the South side of the Brook called Bound Brook.

Item. I give unto my daughter Elizabeth Rogers (after my wifes de-

*Bound Brook is a little stream, called by the Indians Sapokonish, in Eastham.

cease) all my estate, right, title and interest to the Drift-Fish at Panath†, with all my lands and privileges at Easton Harbor, so far as the purchased lines goes during her life, and after her decease, to fall to her youngest son then living by John Rogers.

All the rest and residue of my estate, real and personal, goods and chattels, whatsoever, I, do give and bequeath unto my loving wife, my executrix, equally joint with my sons Stephen Twining and William Twining— all equally joynt-executors full and sole to this my last will and testament And I do hereby revoke and disanulle & make void all other or former wills and testaments by me heretofore made.

In witness whereof I ye said William Twining to this my last will and testament have hereunto set my hand and seal ye 26th day of ye fourth month in the year 1697.

WILLIAM TWINING.

Signed and delivered in ye
presence of
ANN X SCAIFE
JONATHAN SCAIFE.

Then personally appeared —— —— Twining and Stephen Twining (William Twining the other executor being beyond the sea) and proved the above *will*, in due form of law, and probate was to them granted under the seal of the office for probate of Wills, &c. JOHN EVANS.

To the honorable John Evans, Esq., Lieutenent Govenor of Penna and the territories annexed, or to his lawful ordinary for the probate of wills.

Whereas there is a will of William Twining, of Newtown, in our county of Bucks, which said will the executors witnesses being somewhat difficult to be got to Philadelphia through age and other infirmities, they desired me to attest & examine the said witnesses which accordingly I did (viz) the witnesses being legally attested and examined before me, did evidence that they did see and hear the said Wm. Twining seal and acknowledge & declare this said will bearing date the 26th 4 mo 1697 to be his last will and testament,

the 6th of 2d mo. called April 1705.

JOSEPH KIRKBRIDE, Justice."

The above will was found in the Philadelphia Register's office, in the Autumn of 1885, where it had evidently lain 180 years, untouched by the hand of a descendant.

†The word "Panath" in will is wrongly given. It should be "Pamet" or "Paomet." Indian name for Truro, Mass.

ISSUE.

1.—**Eliza,** m. Aug. 19, 1669; Jno. Rogers, son of Lieut. Joseph of Eastham. *Issue:*
   1—Samuel, b Nov. 1, 1671; d inf.
   2—John, b Nov. 4, 1672; m Priscilla Hamblin, 1696.
   3—Judah, b Nov. 23, 1677.
   4—Joseph, b Feb. 22, 1679.
   5—Eliza, b ——— 1682.
   6—Eleazer, b May 19, 16--
   7—Mehit, b ——— 1687.
   8—Hannah, b ——— 1689.
   9—Nathaniel, b Oct 3, 1693.
2.—**Anne,** m Oct. 3, 1672, Thomas Bills; d Sept. 1, 1675, leaving two children: *Annie*, b 1673, and *Eliz.*, b 1675.
3.—**Susanna,** b Feb. 25, 1654 (Orleans Rec.); d young.
4.—**Joanna,** or Johanna, b May 30, 1657; m Thomas Bills, husband of *Anne*, March 2, 1676, and had six children. Left E. and went to Yarmouth, and probably from there to Pennsylvania.
5.—**Mahitable,** supposed to have married Daniel Doane of Eastham, who moved to Bucks county, Pa., 1696, with recommendations from the Sandwich Society of Friends. [See Doan Genealogy of Bucks county, Pa.]
6.—**3 Stephen,** b Feb. 6, 1659; m Abigail Young.
7.—**4 William,** Orleans Records say he was born Feb. 28, 1654; m Ruth Cole.

THIRD GENERATION.

3 **Stephen Twining,** son of **2** William, b. Eastham, Feb. 6, 1659; d. Feb. 18, 1720, Newtown, Pa.; m. Abigail (daughter of John and Abigail Young of E., m. in Ply 12–13, 1648, and had 11 chil.) Jan. 13, 1682-3; d. April 9, 1715, N.

Came to Newtown Tp., Bucks Co., Pa., with his family and his father and mother, 1695. Owned 800 acres of land in Bucks Co., besides considerable property in Eastham. Was appointed Overseer and Elder in Society of Friends, 3 mo. 7-1713, and 2-12-1715, respectively. The Mo. M't'g Records show him to have been an active and leading member of the Society, his name appearing very frequently on the meeting minutes.

The following was entered on the Book of Records belonging to Wrightstown M. M't'g, 11 mo. 3-1776, by Joseph Chapman, Clerk:

"Stephen Twining (son of Stephen) b. 30th, 12 mo. 1684.
    Eleazer   "   "   b. 26th, 11 " 1686.
    Nathaniel "   "   b. 27th, 3 " 1689.
    Mercy   " dau. "   b. 8th, 9 " 1690.
    John   "   "   b. 5th, 3 " 1693."

"The above is a true copy taken out of the Book of Records in Eastham in the County of Barnstable, in the Province of Mass. Bay, in New England, May 31, 1715, by

                        JOHN PAINE,
                          Town Clerk."

Davis' Hist. of Bucks Co., Pa., has the following in regard to Stephen Twining:

"The five hundred acres of Thos. Rowland, extending from Newtown Creek to Neshamony, including the ground the Presby. Church stands upon, was owned by Henry Baker in 1691, who conveyed 248 acres to Job Bunting in June, 1692, and in Oct., 1697, to Stephen Wilson.

1695 Bunting conveyed his acres to Stephen Twining, and in 12 mo 17-1698, Wilson did the same, and Twining now owned the whole tract." In 1703 Stephen owned 690 acres in Newtown. The land on the East of N. was conveyed by S. to his son Eleazer. It is now all gone out of the name and has a large fine residence on the spot where Stephen sat down in his humble log cabin on his first landing on the shores of Penn. In 1707 Stephen again purchased another tract of land, 300 acres, of John Ward, in the Western part of N. Tp. adjoining Wrightstown, which he left his sons Stephen and John, a part of which is still in the name of Aaron Twining, great grandson of 7 John. The Monthly Meetings Rec. tell us that Stephen and Wm. Twining's names were added to a Testimony against selling rum or strong drink to the Indians, dated 4 mo., 3rd, 1699. The same records also show that Friends held meetings at William's house in N., and after his death they were held at Stephen's.

———————WILL——————————A true copy.

I, Stephen Twining of Newtowne, in the county of Bucks, in the province of pensilvania, being indisposed of body but of sound disposeing minde and memory, praised be god for the same, and calling to minde the uncertainty of this life, I do make and ordaine this my present last will and testament in manner and forme as following (viz):

First, my will and minde is that all my just debts and funarall charges be paid and discharged.

Item. I give and bequeath unto my sone, John Twining, the moiety or one-half of a tract of land laying in Newtowne aforesd, purchased from John Ward, the Southwest side thereof adjoining unto land that was formerly Michael Hughs; with all the privileges and appurttenances their unto belonging. To holde to him, his heirs and assigns forever. The other half of the said tract being before conveyed to my sone Stephen twining, and also I give unto said sone John Twining all my *lands or tenements left to me in the government of New England with the appurtanances to him, his heirs and assigns forever.

Item. Whereas, I have already conveyed one hundred and fifty acres, part of the tract of land in Newtowne aforesaid whereon I now Dwell unto my son Eleazer Twining, I do give and bequeath unto my sone Nathaniel Twining two hundred and fifty acres of the tract of land I now live upon, he to have the Northeast part adjoining William Buckman's land and to extend southward downe Newtowne Creek until it come to fifteen pole or pearch on the northside of the house wherein Joseph Lupton did formerly live, and thence such a course or courses as will make two hundred and thirty acres. I give unto the sd Nathaniel out of my low lands and meadows laying by Neshambany Creek, which is now improved (comonly called the Lower Meadows), with all the buildings, orchard and improvements, to hold the same premises with the appurtanances to him and his heirs and assignes forever.

Item. I give unto my two sons, Stephen Twining and John Twining,

* The lands and tenements referred to were bequests of Josiah Cooke's will made 1673 and on record at Plymouth, Mass. He was the stepfather of 2 Wm's wife Elizabeth, and remembered in his will both of her sons, 3 Stephen and 4 William.

one hundred acres of land out of the remaining part of the tract of land which I now live upon to be equally divided between them, to hold to them, ther heirs and assigns forever.

Item. I give unto my Daughter, Rachell Twining, one feather bed with cloathes and curtains and furniture and one larg bible.

Item. I give unto my two Daughters, Mercy Lupton and Rachell Twining, each of them Thirty Pounds current Lawful money of province of pensilvania, to be payed by my executors hereafter to be mentioned out of my personal estate.

Item. All the rest and resedew of my parsonal estate, goods, chattels, Rights and credits whatsover I give and bequeath unto my three sons, Stephen Twining, Nathaniel Twining and John Twining, to be enjoyed by them, their heirs and assigns forever, to be equally divided amongst them.

And lastly, I do hereby constitute, nominate and appoynt my two sons, Stephen Twining and Nathaniel Twining joynt executors of this my last will and testament. And I do hereby Revoke, disannull and make void all former wills and testaments by me heretofore made.

In witness whereof, I, the said Stephen Twining, to this my last will and testament above and within written haing sett my hand and seale the Twentyeth Day of the Twelveth month Anno Domini One Thousand Seven hundred and nineteen (or twenty).

STEPHEN TWINING.

ISSUE.

1.—**5 Stephen,** b. Dec. 30, 1684; m. Margaret Mitchell.
2.—**Eleazer,** b. Nov. 26, 1686,; d. Dec. 17, 1716 N.; m. Jane Naylor, dau. of Jno. of Southampton, Oct. 26, 1716, wid. m. Feb. 4, 1725, Ben Scott of Abington, Pa,
3.—**6 Nathaniel,** b. Mch. 27, 1689; m. Joan Penquite.
4.—**Mercy,** b. Sept. 8, 1690; m. Joseph Lupton July 10, 1713.

Issue:
1. William, b. Jan. 4, 1714.
2. Sarah, b. June 22, 1716.
3. Joseph, b. Jan. 5, 1718.
4. Elizabeth, b. Feb. 30, 1722.

} probably d. young.

After Mercy d. Joseph m. Mary Pickens (wid) July 1730.

5.—**7 John,** b. Mch. 5, 1692–3; m. Elizabeth Kirk.
6.—**Rachel,** m. John Penquite Jr. (2nd wife) 1721 who d. June 28, 1750; she d. Dec. 28, 1779, and was probably born on the eve of her father's coming to Penn.

ISSUE:

1. Jane, b. Sept. 25, 1728; m. Wm. Chapman 1745.
2. Abigail, b. Jan. 1, 1726.
3. Mercy, b. June 19, 1730; m. Ralph Smith, who after his wife's death moved with his children to New Garden, North Carolina.
4. Sarah, b. Jan. 10, 1732; m. Wm. Linton 1766.

7.—**Joseph**, b. Mch. 18, 1696; d. Sept. 12, 1719.
8.—**David**, b. ——; d. July 23, 1711.
9.—**William**,   d. Sept. 9, 1716.

4 William Twining, son of **2** William. Orleans Records say he was b. " Feb. 28, 1654," but probably not until about 1662; d. Eastham, Jan. 23, 1734; m. Ruth Cole, dau. of *John, Mch. 26, 1689, b. 1668.

Very little is known of this steadfast Puritan. His reasons for remaining on the Cape, when his parents, brother Stephen and probably two sisters had gone to Penn., may evidently be found in his unwillingness to compromise the ancestral religion; and he may have been content with that " barren waste " which had produced so many noble men and women, who had truly blessed the world for having lived. Whatever his motives were for not leaving, it remains to his credit that he was a man who transmitted to his descendants those qualities which has made their name an honor wherever they have gone. Not a few of them have filled professions of distinction and the cause of Christianity has been greatly blessed by their devotions. So far as character or intelligence is considered, the reader will determine for himself, as fancy may suggest, while climbing the " genealogical tree," marks of superiority existing between the two great branches. It is certain that the Penn family has

*John Cole, was son of Daniel one of the first settlers of Eastham, who d. Dec. 21, 1694, had eight children.

largely outran the New England branch in point of numbers. The will of his father gives him all the lands or residue lying within the township of Eastham, lying on the south side of Bound Brook and makes him joint executor of said will. At the time the will was probated William is said to have been " beyond the sea." Just what this means is not clear, possibly he may have visited England. His will was made April 13, 1725, and proved Feb. 11, 1734-5.

ISSUE:

1.—**Eliza,** b. Aug. 25, 1690.
2.—**Thankful,** b. Jan. 11, 1697.
3.—**Ruth,** b. Aug. 27, 1699.
4.—**Hannah,** b. April 2, 1702.
5.—**8 William,** b. Sept. 2, 1704; m. Alphia Lewis.
6.—**9 Barnabas,** b. Sept. 29, 1705; m. Hannah Sweet.
7.—**Mercy,** b. Feb. 20, 1708.

FOURTH GENERATION.

5 **Stephen Twining,** son of **3** Stephen, b. E. Dec. 30, 1684; d. June 28, 1772, Newtown, Pa.; m. Oct. 1709, Margaret Mitchell; d. July 9, 1784, in the 99th year of her age. Farmer: will made May 8, 1771; her will 1779.

ISSUE :

All born and lived in Bucks Co. Pa.

1. William, d. July 15, 1711.
2. Elizabeth, b. April 30, 1712; m. Isaac Kirk, Mch. 11, 1730.
    Issue:
    1. Mary, b. Sept. 1, 1730; d. Sept. 9, 1735.
    2. Isaac, b. Aug. 15, 1733.
    3. Joseph, (?) m. Patience Doan 1760.
3.—**Abigail,** b. Dec. 24, 1714; m. Samuel Hillborn and d. before her father, leaving a large family.
4.—**10 Stephen,** b. Feb. 20, 1719; m. Sarah Jansey.
5.—**Mary,** m. John Chapman Oct. 8, 1739.
6.—**William,** b. April 7, 1722 and d. young.
7.—**Margaret,** m. Thomas Hamilton and had a large family.

6 **Nathaniel Twining,** son of **3** Stephen. b. E. Mch. 27, 1689; d. Newtown Tp. Bucks Co. Pa., about 1742, farmer; m. *first* Joan, dau. of John and Agnes Penquite, Oct. 22, 1719, b. Jan. 18, 1696, d. July 27, 1720; m. *secondly,* Sarah Kirk, July 23, 1723, (dau. of John Kirk, who came from Darbyshire England, and settled in Darby, Delaware Co., Pa., 1687, and m. Joan Elliott Jan. 14, 1688, of Kingsessing, at Darby, M. M. and had 10 chil.)

We very much regret having so little regarding our ancestor in direct line. The omission of his death, in the M. M. Records, conveys the idea that he had been previously disowned. By his father's will, he was liberally remembered. A power of attorney, signed by him May 6th, 1742, in which his " loving brother, (7) John Twining of Newtown Tp," is made sole executor of his business, etc., is yet in a good state of preservation, in possession of the compiler. It is a nicely executed document in which the name of Nathaniel Twining is signed in a bold but legible hand.

ISSUE:

1.—**Isaac**, b. May 25, 1724; d. young.
2.—**11 Samuel**, b. Jan. 24, 1726; m. Mary Jenks.
3.—**12 Benjamin**, b. June 3, 1728; m. ——
4.—**Nathaniel, Jr.**, b. Jan. 25, 1730; disowned by Friends 1758.

No trace of Nathaniel, Jr., can be found at all; he probably died without issue.

7 John Twining, son of **3** Stephen, b. Mch. 5, 1692–3, E.; d. Newtown, Pa., Aug. 21, 1775; m. Elizabeth Kirk, Nov. 1718, sister of his brother **6** Nathaniel's wife. She was b. Mch. 9, 1696; d. Nov. 8, 1774, N. Farmer; member Society Friends; Will made Mch. 31, 1773. His descendants are more numerous than any of his cotemporaries.

ISSUE:

1.—**13 John**, b. Aug. 20, 1719; m. Sarah Daws.
2.—**Joseph**, b. Jan. 11, 1720; d. Dec. 28, 1733, Newtown, Pa.
3.—**David**, b. June 17, 1722; d. Dec. 2, 1791 (?); m. Eliz. Lewis about 1762. Was a prominent man in society and business; served two terms in the Penn. legislature; raised Edward Hicks, a prominent Quaker preacher. Will made Oct. 25, 1691. *Issue:*

THE TWINING FAMILY. 37

1. Elizabeth, m. William Hopkins of Phila. (lawyer), who d. July 12, 1803, and had one child, who m. Richard Loines of New York City.
2. Beulah, b. 1770; m. Dr. Torbert, from whom she was divorced.
3. Sarah, m. Thomas Hutchinson and had David, Thomas and Eliz.
4. Mary, m. Jessie Leedom, May 23, 1788, who d. March 21, 1845.

4.—**14 Eleazer**, b. June 8, 1724; m. Mary ——

5.—**William**, b. May 25. 1726; d. Sept. 13, 1814, Bucks County; single.

6.—**Thomas**, b. June 28, 1728; d. Jan. 5, 1733.

7.—**15 Jacob**, b. Oct. 25, 1730; m. Sarah Miller.

8.—**Rachel**, b. Nov. 11, 1732; d. Dec. 22, 1733.

9.—**16 Stephen**, b. April 5, 1734; m. Mary Wilkinson.

8 William Twining, son of **4** William; b. Eastham, Sept. 2, 1704; d. *Orleans, Nov. 17, 1769; m. Apphia Lewis, Feb. 21, 1728. Will made Mch. 19, 1769; proved Dec. 12, same year. In it he mentions his wife Apphia; sons, *Thomas* and *Elijah*, and grand daughter Apphia Rogers. Wid. was member of Orleans Church (Cong.) before 1773. It is said by one of his des. that he was a lawyer in O., but of this we have no proof.

* Orleans was set off from Eastham in 1797. Since that date South Eastham has been known as Orleans, some portions of which is the oldest part of Old Eastham.

ISSUE:

1.—† **Abigail**, b. Dec. 28, 1730.
2.—**17 Thomas**, b. July 5, 1733; m. Alice Mayo.
3.—**Ruth**, b. Dec. 30, 1736.
4.—**William**, b. March 16, 1739; d. 1759.
5.—**18 Elijah**, b. Nov. 4, 1741; m. Loise Rogers.
6.—**Eleazer**, b. about 1746; d. 1762, aged 16 years.

9 Barnabas Twining, son of **4** William, b. Sept. 29, 1705, Eastham; d. "March 5, 1766, aged 60 years," says his tombstone at Orleans; m. Hannah Sweet, June 11, 1731; Admn. of estate granted April 16, 1766, at which time mention in will is made of wife Hannah; sons, Jonathan, Barnabas, Prince and Timothy and daughter Hannah. Widow evidently died at Eastham, 1793, and was member of § Orleans Church before 1773.

ISSUE:

1.—**19 Jonathan**, b. March 26, 1732; m. Tabitha Higgins.
2.—**20 Barnabas**, b. July 7, 1737; m. Abigail Nickerson.
3.—**John**, b. Dec. 9, 1739; probably d. young.
4.—**Stephen**, b. March 19, 1742; d. young.
5.—**21 Prince**, b. July 23, 1744; m. Hannah Rogers.
6.—**Hannah**, no record.
7.—**Timothy**, d. 1777 (?).

† One of the daughters of **8** William Twining, m. a son of Cripp Rogers, of Eastham. A dau., Apphia, b. about 1751, m. Dec. 24, 1772, Ebenezer Harding of E, who moved to Tolland 1783; d. Feb. 14, 1832. Orlow E. Harding, a son, was b. 1807; res. Tolland, Mass.

§ Orleans Church Records, 1774, mention *Bathsheba Twining*, unm., as being member of the church. To which of the families she belonged is not known; possibly she is entered by another name.

FIFTH GENERATION.

10 Stephen Twining, son of **5** Stephen, b. Feb 20, 1719, Bucks Co., Pa.; d. Sept. 3, 1777, B. In 1761, Aug. 14, he bought a farm of 118 acres of Jno. Kerboah, in Springfield, Pa., for £ 601, where he probably for a time lived; m. Sarah Jansey (wid. with two chil.) 1773. The wid. whose maiden name was Worth, m. James Bunson, 1782, son of Joseph of Springfield.

ISSUE:

1.—**Mary**, b. 1774; m. Abraham Wilkenson, March 11, 1705.(?)
2.—**22 Stephen**, b. 1776; m. Elizabeth Baldwin.

11 Samuel Twining, son of **6** Nathaniel; b. Jan. 24, 1726, Newtown, Pa., on the large and productive farm now (1885) owned by Joseph Klette, just outside of the Borough limits. The farm north and adjoining was owned by his brother, **12** Benjaman. m. *Mary Jenks, daughter of Thomas and Mercy (Wildman) Oct. 26, 1752. She was b. April 20, 1733; d. 1803. The Middletown M. M. Records has the following relating to their marriage:

"At M. Mo M't'g 14th of 9mo 1752 Samuel Twining & Mary Jenks, declared their intentions of marriage for the 'first time' & at the M't'g held the 5th of 10mo, they appeared the '2d time', signifying their intentions of m. with each other, & the said Samuel having produced a certificate from Wrightstown Mo. M't'g, to the satisfaction of Friends, they are left at liberty to consumate their said intentions when they see convenient."

The Comm. appt'd for the purpose reported to the meeting of 11mo 2d 1752 that the marriage was "decently accomplished. on 26 day of last mo."

For some trivial offense he was disowned from among Friends July 8, 1766, and therefore the date of his death is not mentioned in the above named records.

ISSUE:

1.—**32 Thomas**, b. Aug, 20, 1753; m. Sarah Crook.

2.—**Samuel**, b. Nov. 16, 1754. Testified against May 2, 1776, for serving with the militia; was a smart man for business, but had spells of being insane; killed by falling from an upper story in a flouring mill; unmarried.

3.—**Isaac**, b. Oct. 13, 1756, and probably d. young.

4.—**Nathaniel**, disowned Dec. 6, 1792, for going into the army; d. unmarried.

5.—**Sarah**, b. May 21, 1758; m. Isaac Van Horn; disowned Jan. 5, 1792; supposed without issue. She may have been his second wife. (See 55 Jacob Twining.)

6.—**Mercy**, m. Dec. 20, 1802, by Isaac Hicks, Edward Bradfield; children: *Mary and Ann*.

7.—**Mary**, m. Stephen Field, May 12, 1792.
Issue: *Elizabeth, Mary and Mercy*.

8.—**Ann**, d. unm.

9.—**Elizabeth**, m. Moses Winner; disowned Jan. 5, 1792.
Issue, several children whose names are not given.

10.—**Joseph**, b. July 31, 1764; d. July 10, 1811; buried Middletown, Pa.

11.—**24 John**, b. Jan. 1, 1761, as given in his Bible, but the M. M. Records say Dec. 31, 1759; m. Becca Bennet.

*The Jenks family in Am. date back to 1700, in the person of Thomas (son of Rev. Ben. Jenks of Eng.), who d. while preparing to emigrate to Am. with his wife Susan and son Thomas, Jr. Thos., Jr., b. 12mo 1699 o. s.; m. Mercy Wildman, dau. of Jno. and Martha, of Middletown, Bucks Co., Pa., 1731, who died 1787; he d. 1797, interred in M. *Issue:* 1 —Mary who m. Samuel Twining; 2.—John; 3.—Thomas, Penn. member of the Constitutional Con. and Senator, d. 1799; m. Rebecca Richardson 1762; 4.—Joseph; 5.—Elizabeth; 6.—Ann. They were members of Friends in good standing, and with the Twinings were among the first people of Bucks Co. of their day.

12 **Benjamin Twining,** son of **6** Nathaniel, b. June 3, 1728, Wrightstown; d. evidently in Warren county, N.J. Very little has been gleaned respecting Benj. The M. M. records show that he was disowned from among Friends 1758. Sold his farm of 53 acres near N., 1757, to John Harris, merchant. His house was the headquarters of Gen. Washington whilst in Newtown, immediately after the battle of Trenton in Dec., 1776. Was a tailor by trade. It has not been learned to whom he m. or the date thereof.

ISSUE:

1.—**Elizabeth**, lived in Chester county, Pa.
2.—**Mary**, lived in Bucks county, Pa.
3.—**Hester** lived in Oxford tp., Warren county, Pa.
4.—**25 Daniel**, b. about 1776; m. Hannah Snyder.

13 **John Twining,** son of **7** John, b. Aug. 20, 1719; d. 1792-3, Bucks county, Pa.; m. Sarah Daws, July 3, 1743; d. 1806. His will made May 21, 1791; her will made April 2, 1805.

ISSUE:

1.—**26 Joseph**, b. Oct. 14, 1748; m. Mary Lee.
2.—**Rachel**, b. Aug. 15, 1751; d. Dec. 28, 1777; m. April 4 1770, Timothy Balderston, who d. May 14, 1827, aged 81 *Issue:* John, David, Mary, Lydia, Timothy and Isaiah all of Bucks county.
3.—**Elizabeth**, who m. Joseph Briggs, of Bucks county; she had five children and d. 1816.
4.—**Mary**, who m. a Tomlinson.

14 **Eleazer Twining,** son of **7** John, b. June 8, 1724; d. about 1801; m. Mary ——, who d. April 17, 1790; all of Bucks county; will made July 8, 1798. Friends.

ISSUE:

1.—**Malhon,** b. March 25, 1761; d. Dec. 6, 1786.
2.—**Hannah,** b. Dec. 21, 1762; d. June 23, 1815; m. Robert McDowell, who d. Nov. 30, 1838, Bucks. *Issue:* 1 William, 2 Mary, 3 Ann, and 4 Eleazer T., who was a a distinguished lawyer of Bucks county; he was received to Friends Feb. 4, 1800.
3.—**27 Silas,** b. Feb. 13, 1765; m. Eliz. Welding.
4.—**Ann,** b. Feb. 21, 1767; m. 56 John Twining; d. Dec. 5, 1815.
5.—**28 David,** b. May 10, 1769; m. Martha Tucker.
6.—**Eleazer,** b. Nov. 13, 1771; d. Dec. 21, 1789.
7.—**Mary,** b. Feb. 20, 1774.

15 **Jacob Twining,** son of **7** John, b. Oct. 25, 1730; d. Oct. 6, 1804, Wrightstown; m. Sarah, (dau. of Henry and Susanna Miller, of Vir. He came from Saxony about 1750; educated for the ministry; she was born at Tolock, on the Rhine,) June 5, 1781; b. Sept. 16, 1757; d. Jan. 10, 1845, Bucks county; Friends; will made Sept. 11, 1804.

ISSUE.

1.—**Elizabeth,** b. Mar. 21, 1782; d. May 23, 1849, N.; single.
2.—**29 John,** b. Aug. 11, 1783; m. Sarah Harding.
3.—**Sarah,** b. Nov. 5, 1784; d. Oct. 8, 1875, N.; single.
4.—**30 Jacob,** b. June 30, 1786; m. Priscilla Buckman.

5.—**Mahlon**, b. Nov. 20, 1787; d. Oct. 11, 1789.

6.—**Susanna**, b. Jan. 22, 1789; d. April, 16, 1882, W.; single.
"A devoted Friend, a remarkable woman."

7.—**31 David**, b. Feb. 5, 1791; m. Hannah Taylor.

8.—**William**, b. Feb. 12, 1794; d. Mch. 31, 1794.

9.—**Rachel**, b. March 4, 1796; d. July 10, 1880; single, res. W.

10.—**Hannah**, d. inf.

11.—**Henry Miller**, b. Oct. 17, 1799; d. May 2, 1875, Phila.; buried at W.; m. Anna M. Gilland of Pittsburgh, Mar. 13, 1851; d. Dec. 28, 1886; no issue.

He was a prominent lawyer, teacher, writer, and traveler in Eastern Countries. Extracts from a tribute to his memory by E. P. Jones, a prominent lawyer of Pittsburgh, Penn.

"Although he was not a native of Pittsburgh, still he has spent the greater portion of his useful life amongst us, and very many of his former pupils, who are to be found among the most accomplished and fashionable ladies of this community, will be pained to hear of his demise. He was a fine scholar, a gentleman, and an honest man, the noblest work of God He was an accurate writer of fine cultivated classical taste, as has been fully demonstrated by his many literary productions that have been published during the last twenty-five years.

He was a finished teacher, thorough in all he taught. He was principal for a number of years of one of the best female Academys we ever had in the city, and although he was most strictly thorough with all his scholars, they all loved, honored, and esteemed him * * * * *
* * * * * * * * * * *

He was raised a Quaker, but died a member of the Episcopal Church He died as he had lived, beloved by all who knew him."

> "Leaves have their time to fall
> And flowers to wither at the North-wind's breath,
> And Stars to set—but all, Thou hast all seasons
> For thine own, O Death!"

16 Stephen Twining, son of **7** John, b. Apr. 5, 1734; d. Feb. 5, 1810, Bucks; m. Mary, dau. of Jno. Wilkinson, Esq , Apr. 18, 1765; farmer and Friend.

ISSUE:

1.—**John**, b. Aug. 20, 1767; d. young.
2.—**Elias**. b. Mar. 26, 1769, Bucks; d. Aug. 20, 1832; m. Mary Stokes, Apr. 16, 1694; d. Sept. 27, 1809, aged 37 years. W., farmer; Friend. *Issue:*
   1. Ann, b. Nov. 28, 1795; m. Malichi, son of 55 Jacob Twining. (See again.)
   2. Sarah, b. Dec. 24, 1796.
3.—**Rachel**, b. Aug. 25, 1771; m. David Watson. Their daughter Mary was the mother of Prof. Ed. H. Magill, Pres. of Swarthmore Coll., Pa.
4.—**Thamer**, b. Feb. 10, 1774; m. David Palmer, Nov. 15, 1797, W.; d. Feb. 21, 1808. *Issue:*
   1. Ann, b. Sept. 25, 1798; d. April 17, 1870; single.
   2. Mary, b. Apr. 12, 1800; m. Joseph Rich, Bucks Co.
   3. Mark, b. Jan. 18, 1802; d. Dec. 17, 1873.
   4. David, b. Jan. 10, 1804; d. Jan. 2, 1873.
   5. Tracy, b. Feb. 1, 1806; d. Nov. 26, 1865.
5.—**32 Jacob**, b. Jan. 28, 1776; m. Margery Croasdale.
6.—**Mercy**, b. July 19, 1778; d. minority.
7.—**Elizabeth**, b. Oct. 23, 1780; d. minority.
8.—**Mary**, b. Jan. 10, 1783; d. Sept. 17, 1803.

17 Thomas Twining, son of **8** William, b. E., July 5, 1753; d. April 23, 1816, Tolland, Mass.; m. *first*, Alice Mayo 1755; d. soon after; m. *secondly*, Anna Cole (Doane), Oct. 24, 1765; d. Oct. 12, 1828, aged 87 years He was a corporal in Cap't. Sam. Knowle's company in 1758, in the war between France and the Provinces.   Was a member of Orleans (So. Church) Cong. Ch. before 1773.   Was a carpenter by occupation, as is evident from a contract extant, for building a house for his cousin, 21 Prince Twining of E., dated April 22, 1772.   He also built the residence now owned by Homer P. Twining, of New Boston, Mass.

In 1783 he conveyed his house and lands in Orleans to Simeon Higgins, and with his brother, 18 Elijah, settled in Granville (1810 Tolland), Mass., where they purchased an extensive tract of land upon which they lived the rest of their days.

In 1795 the Cong. meeting house was built, and in 1797 the West Granville (now Tolland) Cong. Church was organized, of which Thomas was chosen deacon, and was afterwards known as "Deacon Thomas Twining."

He and his wife Anna were very puritanical in observing the Sabbath; she was a great Bible reader, and it is said, read in a singing tone, while the droppings from her pipe burned little holes in her dresses. Although his material circumstances were abundant for one of those days, his brother Elijah seems to have possessed the considerably most property.

The grave stones of these two grandly noble brothers yet stand in the Tolland graveyard, where lay many of the children and grandchildren of each

ISSUE:

1.—**33 Stephen**, b. Sept. 28, 1767; m. Almira Catlin.
2.—**34 William**, b. Dec. 14, 1769; m. Rebecca Brown.
3.—**Alice**, b. Feb. 6, 1772; d. 1846–7 N. Y. City, m. James Graham, merchant, b. Dec. 16, 1773; d. 1829 (?) N. Y. City, Presbyterians.

ISSUE:

1. Jane Maria, d. in N. Y. City at her sister's, Mrs. Gardner, single.
2. Ann Eliza, b. Jan. 25, 1798, Catskill, N. Y.; m. Jan. 25, 1822, Timothy Jones, of Otis, Mass., where he was b. 1792; farmer and mill-wright. In 1870 moved to Becket, and in 1873 to Washington, D. C., where he d. 1886; she d. Aug. 26, 1875; *issue*, seven chil.
3. Harriett, d. unm.
4. Almira, m. Stephen Bosworth or Boswick, of Sandisfield, Mass.; four chil.
5. Emeline, b. about 1805; m. James D. Gardner, res. N. Y. City.; four chil.
6. James Henry A., m. Ester Thorp; physician, res. Scarsdale, N. Y.
7. Adeline L., m. Louise Burdsell; went to South America where he d.; she moved to Staten Island and d. about 1884.
8. Adelia, d. before she was thirty years old.
9. Frances, m. Anson G. Cobb, both dead; 3 chil.
10. Julia H., res. Tompkinsville, Staten Island.

4—**Apphia**, b. 1774; d. April 1843; "baptized by Rev. Bascom, of Orleans Ch., June 26, 1774;" m. Chauncey Fowler. *Issue:*
1. Almyra, b. June 1789; d. 1865, Vineland, N. J.; m. Jno. H. Allen of Sandisfield; eight chil.
2. Hannah, b. July 1800; d. Aug. 1850; m. Percival Davison; three chil.
3. Alonson, b. May 26, 1802; m. May 1834 Sarah E. Miller, both living with their only child, Milton, at Poughkeepsie, N. Y.
4. Chauncey B., b. Jan. 1804; d. unm.
5. Percy M., b. Nov. 1806; m. June, 1839, Lois E. Miller, res. Winsted, Conn.; three chil.
6. Apphia, b. March, 1808; m. Alonzo J. Maltbie, res. Vineland, N. J.
7. Anne T., b. Oct. 1810; m. 1836, Pliney S. Miller of N. Y.; d. May 1842, leaving two dau. and one son, living.

5.—**Anne**, b. 1777, baptized, Orleans, June 29, 1777; m. Col. Joseph Wolcott, Oct. 22, 1810, who d. March 23, 1847, aged 72; she d. Dec. 23, 1861; both buried, Sandisfield, one dau. Lois.

6 and 7—. Two children whose births are not given but whose deaths are stated in Rev. Bascom's "Bill of Mortality," in his Parish, Orleans, 1781.

18 Elijah Twining, son of **8** William, b. Nov. 4, 1741, E; d. Oct. 2, 1802, Tolland; m. Oct., 1762-3, Loise or Louis Rogers (dau. of Judah Rogers, who d. 1773, and whose wife was a Nickerson); d. April 30, 1815, aged 71 years.

Came to Granville, Mass., in the spring of 1783; rolled up a log hut and began to clear the land, and in the fall returned for his family. He went down and back from the Cape, three times on foot. On one of these "tramps" it is said he took $6,000 in specie, and on the way called at a tavern where he thrust his money bag in a corner. A darkey in sweeping the floor had occasion to move the money, but to his credit it is said he *only* looked up and a smile encircled his dusky face. Such a union of confidence and recklessness would be hard to find in these days of vigilance. His dinner cost 12½ cents.

Wolves and bear were plenty and could be heard paddling in the brook just at the foot of the hill on which the log house stood. They drove the stacks of grain full of "spikes" to keep out the wolves. At the Cape, their hay was poled upon the uplands and there left to have the rains wash the salt out of them. In consequence of this excessive saltiness, the cattle they raised were small in stature. Usually four pairs of cattle were attached to a plow, and they would go around a piece of land just seven times in a day, so large were the pieces plowed. Besides property at Eastham, Elijah owned about 2,000 acres at Granville. During his residence at E. he was at one time Constable and Collector. He was a very accurate and systematic business man, whose judgment was good and word was law.

Member of Orleans Church (Cong.) before 1773.

It is said he gave more than any other member to his Church.

Deeded his property to his children in 1800.

Thos Twining

Elijah Twining

Autographs of 17 Thomas and 18 Elijah Twining.

ISSUE:

1.—**35 William**, b. Nov. 13, 1763; m. Tabiatha Smith.
2.—**36 Eleazer**, b. May 29, 1765; m. Mercy Smith.

3.—**Ruth**, b. Dec. 2, 1766; m. ——Smith; d. Colebrook, Conn.

4.—**Joseph**, b. Sept. 28, 1768; d. 1773; baptized Oct. 25, 1772.

5.—**37 Judah**, b. Jan. 21, 1774; m. Cathrine Fowler.

6.—**Abigail**, b. 1775; baptized Nov. 12, 1775 and probably d. same year.

7.—**38 Lewis**, b. April 11, 1777; m. Jeannett Smith.

8.—**Timothy**, b. 1782; baptized Feb. 16, 1783; d. Sept. 22, 1824, Tolland; m. Betsey Hall, dau. of Nathan; d. Jan. 4, 1830. Had one dau., Marnida, who m. Austin Goodsell, and moved to Jefferson Co., Ohio.

9.—**Susanna**, b. April 28, 1787, T.; d. Sandisfield; m. Edward Wolcott, brother of Col Joseph W., who m. Anna, dau. of 17 Thomas Twining. *Issue:*

    1. Lois, m. Julius Deming.
    2. Darius, m. Mary Callender, of Sheffield, Mass.

10.—**Lois**, b. Oct. 8, 1790; d. Feb. 14, 1810, Tolland.

19 Jonathan Twining, son of **9** Barnabas; b. M'ch 26, 1732; d. 1812, Orleans, Mass.; m. *first,* Tabitha Higgins, Feb. 28, 1754; d. 1774; m. *secondly,* Sarah Rogers, M'ch 18, 1775. He was member of Orleans Church before 1773, and wife Sarah in 1793.

Will proven 1813, wherein mention is made of wife Sarah; sons *Barnabas* and *Nathan;* dau. Mercy; grandson Solomon Higgins; grand daughter Tabitha Rogers and Meriam Calking Samuels or Samvels.

ISSUE: All b. at Orleans.

1.—**39 Nathan**, b. M'ch 8, 1755; m. Sarah Clayton.

2.—**Lydia**, b. Oct. 7, 1756; d. 1777, O.

3.—**John**, b. Oct. 11, 1758; d. before 1812.
4.—**Abigail**, b. M'ch 20, 1760; d. before 1812.
5.—**Elizabeth**, b. Dec. 14, 1761; d. 1777 (?).
6.—**Tabitha**, b. Sept. 21, 1763; d. before 1812.
7.—**Mercy**, b. April 18, 1765; was living 1812.
8.—**40 Barnabas**, b. May 14, 1767; m. Rebecca Rogers.

20 Barnabas Twining, son of **9** Barnabas, b. July 7, 1737, O.; d. 1829, aged 92 years; m. *first*, Abigail Nickerson, of Harwich, Feb. 18, 1762; d. June 6, 1790; m. *secondly*, Abigail Knowles, Nov. 14, 1790; d. 1805; m. *thirdly*, Mrs. Hannah Smith, Oct. 25, 1806, d. 1824 (?).

These were all members of the O. Church. The old house owned by Barnabas yet stands in Orleans Tp.; the land adjoins the land formerly **17** Thomas Twining's.

ISSUE:

1.—**Polly**, d. about 1789.
2.—**Martha**, b. Dec. 21, 1764; m. Samuel Cole, Nov. 10, 1785, O.
3.—**Hannah**, b. Dec. 7, 1766; m. Zacheus Higgins, M'ch 9, 1786, Orleans, and had a large family.
4.—**Abigail**, b. Oct. 25, 1768; baptized Sept. 1, 1776, O.; m. Jan. 8, 1801, Curtis Hopkins, Orleans. *Rosilla*, a dau., d. 1888, Orleans.
5.—**David**, b. April 11, 1774; bapt. April 17th; m. Cynthia Gould, April 13, 1797, dau. of Dr. Samuel Kendrick. Removed from Orleans to Woonsocket, R. I., about 1812. *Issue:*

    1. Cynthia, b. Dec. 10, 1798; evidently m , but no issue; d. M'ch 24, 1873, W.

2. Abner, b. Jan. 14, 1800; lived most of his life in Cumberland, R. I., and d. between 1860 and 1870; unm.
3. Samuel, b. Nov. 30, 1803; "went away and they never heard from him after he left, and he probably never m."
4. Sabra, b. Nov. 6, 1806; d. Aug. 28, 1881-2; m. —— Hendricks; lived in Woonsocket. Children: *Edward, George, Ellen, Lyman* and *Adelbert*.
5. Rozilla, b. June 26, 1808; m. *first*, —— Whipple, and *secondly* —— Steese; res. Woonsocket (1887). ISSUE: *Marietta, Lucina, Amos, Cyrus, James and David*.

6.—**41 Abner**, b. Jan. 20, 1772, E.; m. Mary Snow.

21 Prince Twining, son of **9** Barnabas; b. July 23, 1744; d. 1825, Orleans; m. Hannah Rogers, Jan. 3, 1771; d. 1826 (?). Deacon of Orleans Church from 1812 to demise. Will proven 1825, in which mention is made of wife *Hannah*, son *Prince* and five daughters

ISSUE:

1.—**Thankful**, b. Aug. 31, 1773, O.; m. first, Joshua Higgins, Jan. 15, 1804 (?), and secondly, Joseph Snow. Moved to Hampden, Penobscot Co., Maine. Chil.: *Read* and *Nathan* and one daughter.
2—**42 Jonathan**, b. M'ch 25, 1775; m. Tamzin Snow.
3.—**Hannah**, b. June 4, 1777; m. Abner Mayo, Jan. 2, 1800, O., four daughters, one living (1886), O.
4.—**Lydia**, b. April 22, 1779; m. James Rogers, M'ch 2, 1797, b. Oct. 20, 1773; she d. Oct. 18, 1859, O. *Issue:*
    1. Elizabeth, b. Jan. 4, 1798; d. M'ch 2, 1883; m. Ben. Wardell.
    2. Viana, b. Nov. 3, 1800; d. M'ch 28, 1868; m. Tim Atwood 1846, O.
    3. Lydia, b. Feb. 21, 1803; d. Nov. 12, 1879; m. —— M'ch 25 1825.
    4. Ruth, b. Jan. 17, 1805; d. Oct. 4, 1805.
    5. Davis, b. M'ch 15, 1807; d. May 30, 1861.

6. Ruth, b. Aug. 12, 1810; d. Nov. 7, 1879.
7. Ashel, b. July 18, 1812; d. Jan. 11, 1814.
8. Hannah, b. April 21, 1815; d. inf.
9. Lucy, b. M'ch 11, 1816; d. inf.
10. James, b. M'ch 12, 1818; res. Eastham, Mass.
11. Ben., b. M'ch 22, 1821.

5.—**Cloe**, b. April 18, 1781; m. Joseph Atwood, Sept. 17, 1810, O.; no issue.

6.—**43 Prince**, b. April 30, 1783; m. Mary Higgins.

7.—**Lucy**, b. April 29, 1785; d. of cancer, unm., about 1848; raised her nephew **89** Jonathan Twining.

SIXTH GENERATION.

22 Stephen Twining, son of **10** Stephen; b. 1776, Newtown, Pa.; d. about 1848, Bucks Co. After reaching his majority, went among the Indians in Cattaraugus Reservation, N. Y., where he spent nine years as teacher and counselor, with the approval of Friends of Phila. Yearly Meeting. About 1818, he m. Elizabeth Baldwin, a prominent Friend Minister of Troy, N. Y. She commenced preaching when about 19 years of age and traveled much before and after m.; was an invalid for many years; author of religious works. After his wife's decease, 1827, Stephen returned from Troy to Bucks Co.

ISSUE:

1.—**44** Charles, b. Troy, Aug. 9, 1820; m. Eliz. West.
2.—Sarah, m. Isaac Simpson, son of Jno.; moved from Bucks to Independence, Kan., where he died. They had two sons, John and Willet, both dead; wid. res. (1887) in Independence.

23 Thomas Twining, son of **11** Samuel; b. Aug. 20, 1753, Wrightstown; d. Jan. 29, 1838, North Boston (Podunk), Erie Co., N. Y.; m. Sarah, dau. of Samuel Crook,* Sept. 27, 1781.

* Samuel Crook was des. from John Crook, a celebrated Quaker preacher and writer, b. 1617, d. 2-26-1699, at Sewel, in Bedford, Eng., where he was b. (Friend's Lib. Vol. 12, 1849.)

Nathan C., a son of Samuel, was killed at the battle of Lake Erie, on the flag ship Lawrence.

Gen. Crook of the U. S. Army is said to be closely related to this *Samuel*.

Moved to Quakertown, N. J., where he took a certificate of membership from the Middletown, Pa., to the Kingswood, N. J., Mo. M't'g (now Quakertown), dated 1st day, 2nd mo., 1781, wherein he is said to be free from marriage engagements and a faithful attendant of Friends' Meetings.

He purchased July 3, 1793, a farm of 140 acres, of "James Parker and Gertrude, his wife," for £388, within two miles of Q.

On this beautiful *farm he erected a grist and fulling mill. Was a clothier by trade and owned considerable property for one in his day, being worth $16,000, all gained by his own industry. Refusing to bear arms, he was called a Tory, was shot at during the Revolution and compelled to flee.

* "The original Twining property contained 140 acres, mostly timber. The grist mill (stone) was built by "Tommy;" it is in good condition, and rents for $500 dollars. The old stone buildings across the way was Tommy's Fulling Mill. It is still in good condition and is used as a dwelling."

"The old Twining house stands as firm as a rock, and with the very extensive additions (built since), it is one of the most imposing farm houses in this region. The old "grandfather's clock," which was purchased at Tommy Twining's vendue, still stands in the corner of the room where it stood 90 years ago. Was probably made in England, and is in good condition, both in body and mind, and will doubtless number the days and hours for many generations to come. Above the dial the Crescent moon was visible, indicating that dark nights will soon be followed by moonlight nights, balmy breezes, frog concerts, and Spring's fresh budding hours."

"People wondered that Tommy T. wanted to sell such a valuable property. But he was anxious to move to the 'Holland Purchase.' The property is very valuable. Several hundred acres have been added to the original purchase. An oil mill has been built on the creek below the grist mill, besides several dwellings and shops."

"Quakertown is now an old place, but when the first settlement was made is not known. There is one old homestead in the Wilson family with a stone house that bears the date 1734, on its gable: the walls are three feet thick."

"There was a Quaker Church built in Q. in 1744, on the site of an old

In 1811 he moved with his family to North Boston, Erie Co., N. Y., where he purchased 500 acres of land, and was engaged the remainder of his life in farming.

The Kingwood M. M. granted him, his wife Sarah and two children, Thomas and Selinda, certificates of removal to Pelham, Upper Canada, (prob. nearest M. M.), dated Sept. 12, 1811. John and Charles, sons, were granted similar letters of same place and date.

He was a tall man, heavy set, very pleasant in appearance and very much of a business turn. Wife, Sarah, d. Mch 28, 1841. Her father done business with Thomas at Q.

ISSUE:

1.—**Samuel**, b. Oct. 4, 1782; d. April 5, 1785, Q.
2.—**45 John**, b. Dec. 2, 1784; m. Sally Palmer.
3.—**Rachel**, b. June 1, 1787; d. Aug. 4, 1866, Brant, Erie Co., N. Y.; m. *first*, David Laing, N. J.; b. Jan. 3, 1782; d. Dec. 24, 1821, Buffalo, N. Y.; m. secondly, —— Widdifield.

meeting house that was consumed by fire. I well remember attending the meeting in my younger days and the countenances of the old patriarchs, who sat in the "gallery," are still fresh in my memory. It was an antiquated structure with double roof and immense chimney, and afforded quarters for soldiers during the Revolution, and there were many marks on the floor, caused by their Camp Kettles being removed from the immense fire place where the cooking was done."

"The Graveyard is enclosed by a moss grown wall which has stood for nearly two centuries. It is known as the 'Old Quaker buying ground of Revolutionary Memory.' In a corner obscure and alone is a grave with a marble tombstone and the letters almost obliterated. It is the grave of 'Morris Robeson Esq.,' grandfather of the late Sec. of the Navy under Grant." (John Laing of Quakertown, N. J.)

"A railroad is now (1890) in course of construction through the property and it is probable that the iron horse will invade the old hamlet during the coming summer and it may have a boom and its long Rip Van Winkle sleep come to an end."

THE TWINING FAMILY.

Issue by First Husband:

1. Thomas, b. Dec. 1, 1808, Boston, Erie Co., N. Y.
2. Hugh, b. Dec. 13, 1810; res. Eden Centre, Erie Co., N. Y.
3. William, b. April 11, 1813; d. May, 1874, Eden, N. Y.
4. Isaac, b. June 3, 1815, Boston.
5. Abram, b. Aug. 1, 1818, Boston; d. Nov. 17, 1882.
6. James, b. Sept. 8, 1820, Boston; d. Jan. 16, 1885, Buffalo.
7. David, b. Jan. 10, 1822, Buffalo; d. July 15, 1822, Canada.

Issue by second husband: two daughters.

4.—**46 Charles**, b. July 20, 1789; m. Betsey Boutwell.

5.—**Mary**, b. Feb. 17, 1792; d. July 19, 1865; m. Reuben Johnson, about 1811, six years her senior; d. April, 1840, Boston, Erie county, N. Y., where they lived and died; farmers; Quakers. *Issue:*

1. Sally, b. 1811; m. Wm. Clarke; both dead; no issue.
2. William, b. April 17, 1819; m. Cath. Wilson, April, 1842; children, two daughters; res. Plainwell, Mich.
3. David, b. July 31, 1822; m. Dec. 25, 1842, Emeline Walker, d. Aug. 17, 1881; farmer, Boston, N. Y.; wid., res. Lancaster, Wis.; no issue.
4. Mary, b. 1825; m. about 1845 Geo. Fox Pound; both d. Manchester, Iowa; five children.
5. Hugh, b. August, 1828; m. about 1850, Cordelia Sprague; farmer; res. Fredonia, N. Y.; six children.

6.—**Selinda**, b. Nov. 28, 1796; d. March 23, 1839; m. Daniel Webster. Settled in Eden, Erie county, N. Y., about 1814, where they remained until death; Quakers. *Issue:*

1. Hugh, b. Jan. 14, 1816; m. and has two children; done business in Buffalo, N. Y.; in 1887, settled in Pasadena, Los Angeles Co., Cal.; dry goods and clothing; Baptist.
2. Thomas, b. Feb. 8, 1818; d. 1876, Eden.
3. Joseph, b. March 3, 1820; d. Dec. 11, 1876, Eden.
4. Asalich S., b. March 31, 1822; in business, Buffalo, N. Y., since 1852.
5. Mary, b. May 16, 1824; m. Jno. Gifford, farmer, res. Eden.
6. Sarah, } twins, b. March 14, 1827. { former res. Eden.
7. David, } { d.
8. Daniel, b. Aug. 21, 1829; d. March 15, 1879, Eden.
9. Amy, b. Feb. 19, 1832; m. L. Foster. res. Evans, Erie Co., N. Y.

7.—**Adonijah**, b. May 10, 1799; d. inf.

8.—**47 Thomas**, b. Jan. 13, 1801; m. Sarah Kester.

24 John Twining, son of **11** Samuel, b. Jan. 1, 1761, W.; d. March 25, 1849, Union, Broome county, N. Y., aged 88 years, 2 months and 24 days, as given in the family Bible and inscription on his tombstone. Disowned from among Friends, Bucks county, Pa., 4 mo., 4th, 1782, for training with the militia, but it is probable he was subsequently restored, as his home at Union was a resort for Friends and because his des. so declare. Married Becca Bennett Jan. 19, 1786; b. Aug. 8, 1767; d. Feb. 17, 1854. Removed from Bucks county to Quakertown, N. J., where he was engaged in the fulling business, at one time with a "Mr. Stout," and evidently again with his brother **23** Thomas. In the year 1821 he settled at Union, Broome county, N. Y., where he was preceeded by his son William, who had bought a farm there. At this date of removal he evidently lived in Redington tp., Hunterdon Co., N. J. He was a "thick, heavy-set man, and very pleasant looking," and in his latter days, when very feeble, it is said he was often heard, after retiring, to repeat in trembling tones the Lord's Prayer. The old family Bible from which the family records are obtained, is now in possession of Charles Cleveland, of Broome county, N. Y., a connection.

ISSUE:

1.—**Mary**, b. Sept. 30, 1786; d. about 1850, Campville, Tioga county, N. Y.; m. ―― Powers, of Broome county.
2.—**William**, b. Oct. 22, 1788; d. Jan. 19, 1860, Union, N. Y.; fuller, unm.

3.—**48 Thomas**, b. Sept. 4, 1790; m. Elizabeth McKinzie.
4.—**Rachel A.**, b. July 19, 1792; d. April 10, 1867, Ocean county, N. J.; m. Samuel Wardell, 1819. *Issue:*

    1. Eliza A., 2 Rebecca M , 3 Henrietta, 4 Sarah A., 5 Cathrine A., 6 Mary M., 7 Caroline M., 8 Samuel L., 9 Charles H., 10 John H. Sarah A. res. Toms River, N. J.

5.—**49 John**, b. March 25, 1794; m. Dorcas Fonner.
6.—**50 Samuel**, b. Feb. 22, 1796; m. Eliz. Stout.
7.—**51 Benjamin**, b. Nov. 9, 1797; m. Marianna Atkins.
8.—**Sarah**, b. Oct. 16, 1800; d. March 15, 1867, Broome county; m. Joseph Cleveland, January, 1822; d. March 10, 1876; M. E. church. *Issue:*

    1. Rachel A., b. Nov. 14, 1823; m. Isaac Van Demark, 1849.
    2. Charles, b. July 29, 1825; m. Hannah Van Noy, 1866.
    3. Sarah, b. Dec. 1, 1827.
    4. Martha, b. Sept. 14, 1829; m. Ira Packard, 1849.
    5. Joseph N., b. Jan. 24, 1833; m, Mary E. Plain, 1857.
    6. George, b. Feb. 20, 1835; m. Mary Austin, 1857.

9.—**52 Mahlon**, b. March 20, 1802; m. Lucy L. Goodseede.
10.—**Joseph**, b. April 6, 1804; d. without issue. New York.
11.—**Rebecca Ann**, b. March 6, 1807; d. June 22, 1883; m. Richard Lashier, Nov. 13, 1831, b. June 11, 1805; died Jan. 23, 1841; Broome county. *Issue:*

    1. Theodore, b. Oct. 30, 1832; m. Olive Powers, 1863; two chil.
    2. William, b. Aug. 14, 1835.
    3. Doctor Franklin, b. Sept. 29, 1838; m. Luetta J. Hammond, Oct. 10, 1861; res. Hooper, N. Y.

12.—**Henry Clifton**, b. June 11, 1809; d. Aug. 27, 1866, East Smithfield, Pa., where he moved to from Broome county, 1861; died in the field at work, with dropsy of the heart; farmer and stone mason; m. Chloe Kickok, b. Aug. 29, 1810, Rutland, Vt.; d. July 19, 1874, of lung fever; " Christian." *Issue:*

    1. Rebecca M., b. May 9, 1842; m. Alonzo P. Jones, July 31, 1875, and moved to Towanda, Pa., where he d. June 10, 1876; Baptist; widow; lives at T. (1887.) "Brethren."
    2. Sarah G., b. July 10, 1843; d. Aug. 2, 1845.
    3. John H., b. Nov. 2, 1845; killed or missing in battle of The Wilderness, May 6, 1864; had served two years previous and re-enlisted, Corporal Co. C, 2d Reg. U S. sharp shooters, Col. Berdans.
    4. Oliver C., b. May 27, 1847; d. Dec. 27, 1853.

THE TWINING FAMILY.   59

25 Daniel Twining, son of **12** Benjamin; b. about 1776; d. Sept. 1, 1831, Warren Co., N. J., where he lived and was probably born; tailor; m. Hannah Snyder, who d. Dec. 16, 1831, aged 50 years.

ISSUE:

1.—**John**, b. about 1800; d. aged 70; m. Anne Kisbaugh; *no issue.*
2.—**Christean**, d. in Warren Co.; m. William Ribble 1825; chil., *John, Hannah, Anne, Ibby and Susan.*
3.—**Christopher**, went west about 1838, after his m., when he had three children; m. Sarah Lomison. No trace of him or his descendants can be found.
4.—**53 Benjamin**, b. Aug. 30, 1810; m. Eliz. Lance.
5.—**Frederick**, b. Mch. 31, 1815; m. Joan Mettler; b. April 18, 1816; d. Aug. 15, 1884; settled in Vanatta, Licking Co., Ohio, about 1838; lost his mind after his wife's death; cooper by trade. He d. Dec. 16, 1887; res. V. *Issue:*

   1. John, b. Sept. 5, 1833; d. 1835; Warren Co., N. J.
   2. Cyrus, b. Feb. 13, 1835; d. in the Great Rebellion, at Crup's Landing, with fever; member 76 Ohio; single.
   3. Mary Jane, b. Nov. 15, 1837; res. Vanatta.
   4. William D., b. Mch. 11, 1839; d. Oct. 18, 1874; m. Emeline, Wise; Licking Co.; no issue.
   5. Martha, b. Feb. 12, 1841; m. John W. Hass; res. near Utica, Ohio; five chil.
   6. Alice, b. Mch. 20, 1843; d.; no issue.
   7. Hannah, b. Sept. 12, 1845; res, V.
   8. James F., b. Oct. 28, 1848; d. young.
   9. Hattie, b. Dec. 25, 1850; m. Dora Hartman, b. Feb. 18, 1857, res. V.

6.—**Betsey**, d. 1887, Oxford, N. J.; m. Charles Laning, who d. 1877; chil. *Chris., Charles, Steve, Fanny, Martha, Eliz.* and *Hannah.*

7.—**Martha,** b. June 7, 1812; m. Charles Kennedy, who d. Sept. 1878, Slateford, Pa., aged 68; wid., res. with one of her chil. at S. *Issue:*
    1. Daniel, b. 1832.
    2. Jacob, b. 1833.
    3. Ezra, b. 1835.
    4. George and Harrison, b. 1839.
    5. Elizabeth.
    6. Jeremiah, b. 1842.
    7. Harriet, b. 1850.
    8. Hetty, b. 1852.

8.—**54 Jacob,** b. about 1816; m. Sidney Gano.

26 Joseph Twining, son of **13** John; b Oct. 14, 1748, d. Aug. 18, 1821, Warwick, Bucks Co. Pa.; m. *first:* Mary Lee, Dec. 27, 1769; b. Nov. 23, 1750, d. Oct. 13, 1782; m. *secondly:* Hannah Duffil, b. June 23, 1760, d. July 24, 1841. Chil. all lived and d. in Bucks Co.

ISSUE BY FIRST WIFE:

1.—**55 Jacob,** b. Oct. 7, 1770; m. Phebe Tucker.
2.—**Hannah,** b. March 11, 1772; m. —— Tucker; d. Oct. 6, 1815.
3.—**56 John,** b. Oct. 21, 1773; d. May 27, 1827.
4.—**Sarah,** b. Sept. 11, 1775.
5.—**Mary,** b. Nov. 5, 1778; d. Aug 28, 1822 (?)
6.—**57 Joseph,** b. Nov. 8, 1780; m. Mary Tucker.
7.—**William,** b. Nov. 15, 1782; d. prob. young.

    By Second m.
    8. James, b. June 25, 1784; d. Feb. 7, 1876; m. Martha Tomlinson 1815; no chil.
    9. Elizabeth, b. Feb. 23, 1786; d. Nov. 1876; m. Joseph Tomlinson 1804; five chil.
    10. Edward, b. July 27, 1788; d. July 14, 1851, Warwick; m. Margaret Scott; one son, James, d. Aug. 23, 1851.
    11. Mary, b. April 1, 1790; d. 1846; m. John Scott 1812; seven chil.

12. Rachel, b. June 19, 1795; d. May, 1865; m. 1817 Thomas Tomlinson; eight chil.

13. Deborah, b. Dec. 23, 1797; d. Oct. 31, 1880; m. Francis Tomlinson 1822; five chil.

27 Silas Twining, son of **14** Eleazer, b. Feb. 13, 1765; d. Feb. 26, 1827, Warwick, Bucks Co., Pa.; m. Elizabeth Welding, Dec. 3, 1893; b. 1774; d. 1827; farmer, Quakers.

ISSUE:

1.—**Eleazer**, b. May 27, 1794; d. Feb. 1, 1797.
2.—**Mary**, b. Feb. 1, 1796; d. Feb. 4, 1797.
3.—**Ruth**, b. Nov. 31, 1797; d. 1867; m. Isaac Lacy, Oct. 15, 1823, Wrightstown, who d. July 15, 1881, aged 81 years, 24 days. *Issue:*
   1. Silas, b. Oct. 15, 1825; d. Sept. 20, 1827.
   2. Rachel, b. Nov. 15, 1827; m. Birdsall, and lives in Ohio.
   3. Edwin, b. March 27, 1830; single.
   4. Elizabeth, b. Feb. 9, 1836; single, res. W.

4.—**58 Watson**, b. Nov. 20, 1799; m. Margaret Hallowell.
5.—**Ann**, b. July 15, 1801; d. Jan. 6, 1864.
6.—**Alice**, b. Aug. 13, 1803; d. Aug. 24, 1873.
7.—**Letitia**, b. Oct. 25, 1805; d. 1864.
8.—**59 Silas**, b. March 27, 1807; m. Hannah Harrold.
9.—**Elizabeth**, b. July 10, 1809; d. 1852.
10.—**Samuel W.**, b. Dec. 14, 1810; d. Hampton, Ill., 1847, moved to Il. 1840, where he m. Martha T. Welding, 1842, formerly of Bucks Co.; d. at H. March 10, 1886; Quakers, but joined the Cong. church. Several of the Weldings moved to Hampton from Bucks Co., 1842; all now dead. *Issue:*
   1. Henry C., b. Aug. 13, 1842; d. March 2, 1886, H. He was a highly esteemed man; single.
   2. Caroline, d. Sept. 1, 1846; aged 17 months.

11.—**Amos**, b. April 30, 1813; d. Sept. 6, 1822.
12.—**Mary**, b. Nov. 5, 1817; d. Aug. 29, 1822.

28 David Twining, son of **14** Eleazer; b. May 10, 1769, Bucks; d. April 16, 1823, Northampton, Bucks Co.; m. Martha Tucker, Aug. 4, 1794; b. Sept. 1, 1773; d Jan. 9, 1841. Friends, Buckingham M. M.

ISSUE:

1.—**Mahlon**, b. M'ch 8, 1795; d. Feb. 8, 1797.
2.—**60 William**, b. April 13, 1797; m. Rebecca Riley.
3.—**John**, b. M'ch 23, 1799; d. July 2, 1822, Warwick; single.
4.—**Eleazer**, b. Nov. 3, 1800; d. April 9, 1827; single.
5.—**61 Isaac**, b. Aug. 8, 1802; m. Ann L. Hallowell.
6.—**Phebe**, b. Dec. 23, 1804; d. May 3, 1853; m. John D. Alerson, 1847-8, all of Harford Co., Maryland; no issue.
7.—**62 Thomas**, b. Feb. 16, 1808; m. Sarah A. Bean.
8.—**Beulah E.**, b. Dec. 15, 1811; m. Alex. R. Amos, 1853, res. Harford Co., Md., Upper Cross Roads; no issue.

29 John Twining, son of **15** Jacob, b. Aug. 11, 1783; d. Sept. 19, 1853, Phila., Pa.; m Sarah Harding, dau. of Isaac and Phebe, Sept. 29, 1805; b. 1786; d. Oct. 8, 1870.

He settled on the old homestead in Bucks Co., but afterward moved to Phila.

ISSUE:

1.—**63 Jacob**, b. Aug. 6, 1806; m. Rachel Ryan.
2.—**Phebe**, b. Nov. 6, 1808; d. Mch. 20, 1812.
3.—**64 Abbott C.**, b. Nov. 30, 1810; m. Mariah Warner.
4.—**65 Isaac H.**, b. Oct. 21, 1810; m. Phebe Megadegan.
5.—**John**, b. Nov. 29, 1814; d. Sept. 27, 1823.
6.—**Emily**, b. Feb. 20, 1817; m. Robert Getty, May 8, 1835; b. Glasgow, Scotland, July 6, 1811; res. Oxford, Neb.

Issue:
1. Annie C., b. Feb. 25, 1837; Penn.; m. Smith Tuttle, Dec. 10, 1854; b. Oct. 7, 1832; carpenter, res. Cheney, Spokane Co. Washington.

THE TWINING FAMILY. 63

   2. Sarah T., b. Sept. 22, 1839, Penn.; m. Charles H. Golden, Sept. 22, 1860; b. June 9, 1835, Ct., ship caulker; res. Tottenville, Staten Island.
   3. Emily J., b. Nov. 13, 1842, N. Y.; m. John Burtchet, April 19, 1874; b. May 27, 1847, Kentucky; farmer; res. Oxford, Neb.
   4. Julietta, b. May 9, 1847, N. Y.; d. July 5, 1889, Lincoln, Neb.; m. *first*, James Watson, July 10, 1871; divorced, and m. *secondly*, Elmer Sherman, 1879; b. Arcade, N. Y., Sept 19, 1842; machinist.
   5. Rodmond, b. Jan. 11, 1850, N. Y.; brickmason; res. Oxford·

7.—**Elizabeth**, b. May 3, 1819; m. Thomas Laird (son of Hugh and Margaret), July 23, 1835; b. Nov. 15, 1815, Ireland; d. May 10, 1877. Came from Phila. to Wis. about 1848; merchant, Boscobel, Wis. Wid. res. Montfort, Grant Co., Wis. *Issue:*

   1. Sarah, b. July 11, 1839, Delaware; m. twice; res. Norden, Neb., chil.
   2. Margaret, b. Dec. 4, 1841, Delaware; d. May 19, 1871; m. J. H. Lincoln; res. Montfort, Wis.
   3. Nancy, b. Dec. 17, 1844; m. Thos. B. Dewitt; res. Montfort; chil.
   4. Vintona, b. April 24, 1847; school teacher 16 years; res. Boscobel, Wis.
   5. Susan, b. May 28, 1851; m. Robt. Moran; res. Bertrand, Neb.; chil.
   6. Thomas, b. Oct. 17, 1853.
   7. Jessie F., b. July 3, 1857; m. Henry R. Brown, Mch. 17, 1889; Res. Pickering, Mo. She is a school teacher.
   8. John C. F., b. Sept. 28, 1858; res. Perry, Dakota.
   9. William S., b. Feb. 6, 1861; res. Montfort.

8.—**Sarah**, b. Mch. 1, 1823; unm.; res. Southampton, Bucks Co., Pa.

9.—**Ellen**, b. Jan. 10, 1825; m. John Jaquett, son of Asel and Margaret, Mch. 24, 1844. She d. April 16, 1865; lived at one time in Boscobel, Wis. He res. Waverly, So. Dakota. *Issue:*

   1. Mary, m. Henry Wagner, and went from Boscobel, Wis., to Storm Lake, Ia.
   2. Kate, after her mother's death, lived in Bucks Co., Pa.; m. —— Clark.
   3. Miles, res. So. Dak.
   4. Annie, lives with her father.

10.—**Susannah**, b. Sept. 24, 1827; m. Abraham C. Funston, son of Thos. and Hannah, Nov. 5, 1846; res. Philadelphia. *Issue:*
    1. Oliver, m. and has chil.
    2. Sarah, single.
    3. Hannah, single.

11.—**Mary**, b. M'ch 8, 1830; d. Jan. 25, 1846.

30 Jacob Twining, son of **15** Jacob, b. June 30, 1786; d. Feb. 21, 1871, Bucks Co.; m. Priscilla Buckman, Oct. 12, 1808; d. Sept. 26, 1876, aged 90 years; farmer; Quakers. Lived 63 years on a farm in Northampton Tp., inherited from his father.

ISSUE:

1.—**Thomas**, b. Feb. 14, 1810; d. Dec. 10, 1863 (?), Bucks Co.; single.
2.—**Sarah**, b. Dec. 17, 1811; m. Joseph Smith, b. Feb. 10, 1809; d. May 25, 1882. *Issue:*
    1. Thomas, b. M'ch, 1839.
    2. Margaretta, b. Feb. 23, 1841; m. Joseph Michener; chil.
    3. Priscilla A., b. Dec. 26, 1842; m. John T. Pool; chil.
    4. Mary E., b. Dec. 16, 1845; single.
    5. Henrietta, b. Feb. 17, 1848; m. Ed. T. Slack.
    6. Salle, b. M'ch 30, 1851; single.
    7. Rachel, b. June 25, 1853.
3.—**Mary H.**, b. Dec. 25, 1814; m. Thornton Stackhouse, Feb. 19, 1845; b. Sept. 18, 1810. *Issue:*
    1. Anna L., b. May 21, 1845; m. Franklin Hulme; chil.
    2. James, b. May 13, 1848; m. Sadie Lewis; chil.
    3. Henry F., b. May 16, 1850; m. Sydney J. Jackson; chil.
    4. Emma J., b. June 10, 1852; m. I. H. Jones.
    5. Margaretta M., b. Jan. 1, 1855; d. July 21, 1880.
    6. Ella, b. Oct. 24, 1857; d. Oct. 29, 1857.
4.—**66 Jessie B.**, b. Sept. 25, 1817; m. Hannah Beans.
5.—**67 Henry M.**, b. Jan. 4, 1820; m. Eliz. Stackhouse.
6.—**Jane**, b. Nov. 11, 1822; single.
7.—**Priscilla A.**, b. June 29, 1825; d. Feb. 7, 1834.
8.—**68 Cyrus B.**, b. Sept. 25, 1827; m. Sarah A. Atkinson.
9.—**Abraham H.**, b. April 18, 1830; d. April 14, 1832.

THE TWINING FAMILY.

31 David Twining, son of **15** Jacob, b. Feb. 5, 1791; d Oct. 13, 1877, Wrightstown. Lived after m. on the farm left him by his father until within three years of his death. Family all Quakers; m. *first:* Hannah Taylor, 1818; d April 6, 1830; m. *secondly:* Mercy Van Horn, who d. 1872.

ISSUE BY FIRST WIFE:

1.—**69** Amos H., b. May 31, 1820; m. Mary Tomlinson.
2.—**70** George, b. Oct. 24, 1823; m. Anna C. Eberman.
3.—Elizabeth H., b. M'ch 12, 1826; d. Nov. 5, 1886, Wrightstown; m. Edward Atkinson, Feb. 12, 1857; b. July 24, 1823. Farmer and President of Newtown National Bank, Pa. She was an estimable and talented Quaker lady, to whom the compiler of these records is much indebted. Her persevering research brought to light many interesting facts relating to the Bucks Co. Twinings. Living, as she did, surrounded by the shadows of an ancestry extending back two centuries, afforded her special opportunities, which she improved, as only one could who felt a deep and reverential interest in the family name.

ISSUE BY SECOND WIFE:

4.—Abbott A., b. Dec. 16, 1831; d. Dec. 15, 1832.
5.—Frances M., b. Feb. 12, 1834; living in Philadelphia; m. Francis V. Krusen, Dec. 24, 1857; b. M'ch 15, 1832; family all Friends. *Issue:*
    1. Clara A., b. Dec. 12, 1858; m. Dec. 13, 1888, Edward Atkinson (her uncle); res. Wrightstown; ch. Robert Edward; b. Oct. 28, 1889.
    2. Edward A., b. May 1, 1860; physician; res. Philadelphia.
    3. Ellen C., b. May 6, 1862.
    4. Henry A., b. April 7, 1864.
    5. George C., b. July 8, 1868.
    6. Anna G., b. Nov. 13, 1870; d. July 17, 1871.
    7. Maggie T., b. Sept. 13, 1872.
    8. Charles W. (twin), b. Dec. 25, 1874; d. inft.
    9. Thomas (twin), b. Dec. 25, 1874; d. inft.

ELIZABETH H. TWINING.

AUTOGRAPH OF 3 Stephen Twining.

32 Jacob Twining, son of **16** Stephen, b. Jan. 28, 1776; d. Sept. 30, 1863, N.; farmer; Friend. He was persistent in the claim that his "original ancestors" came from Yorkshire, England, and "prided himself on his English blood;" m. Margery Croasdale, April 2, 1802; d. April 5, 1861; ch. all b. in N., Bucks Co., Pa.

ISSUE:

1.—**71 Croasdale**, b. May 1803; m. Mary Kirk.
2.—**72 Stephen**, b. June 25, 1805; m. Sarah A. Warner.
3.—**Elisha W.**, b. Oct. 27, 1808; d. May 26, 1823, N.
4.—**Charles L.**, b. June 30, 1811; d. April 9, 1883, N.; m. Maria Cooper; dau. of Chillian Cooper; farmer; no children.
5.—**Mary Ann**, b. June 16, 1814; m. Eleazer T. Wilkinson, farmer, of Warwick tp.; d. Oct. 23, 1876.
6.—**Isaac C.**, b. April 6, 1819; m. Hannah, dau. of Chillian Cooper (above) 1863; farmer; res. Carverville, Bucks Co. Friends. *Issue:*
    One daughter, Nettie, b. May 25, 1865.
7.—**Aaron**, b Nov. 29, 1821; m. Emily, dau. of Charles Trego, Dec. 17, 1857; b. July 26, 1825; farmer; Friend; res. Wrightstown. Lives on a portion of the original 3 Stephen Twining purchase made in 1696. *Issue:*
    1. Fannie M., b. Sept. 25, 1859; d. Feb. 5, 1884.
    2. Anna H., b. April 25, 1865.
8.—**Deborah, C.**, b. April 9, 1824; m. Charles R. Scarborough, 1857: farmer; res. W. *Issue:*
    1. Annie C., b. Jan. 10, 1859; grad. Pa. State Normal School.
    2. Edward, b. June 28, 1861.

33 Stephen Twining, son of **17** Thomas, b. Sept. 28, 1767, Eastham; grad. Yale, 1795; lawyer, New Haven, Ct.; Steward and Treasurer of Yale College many years; d. 1832 of heart disease; Cong. Church; m. Almira Catlin, Oct. 2, 1800; b. Aug. 24, 1777, Litchfield, Ct., and d. 1846. An anecdote is related of him as follows: "After Stephen, who was much more disposed to work with his head than with his hands, went to Yale College, the old man and William were plowing with a yoke of cattle, one of which was rather inclined to reflection than to action. The old man, quite out of patience, finally exclaimed: ' What can we do with that lazy off ox?' ' Send him to college!' was the prompt reply." His tombstone in the New Haven Cemetery bears the inscription: " He Feared God."

ISSUE:

1.—**73 Alexander A.**, b. July 5, 1801, N. Haven; m. Harriet A. Kinsley.
2.—**Almira**, b. Dec. 31, 1802; d. Dec. 18, 1809.
3.—**74 William**, b. Dec. 9, 1805; m. Margaret E. Johnson.
4.—**Mary Pierce**, b. July 26, 1809; d. March, 1879. She was " an active and controlling influence for many years in the affairs of certain charitable organizations in N. Haven, Ct., a woman of great energy and spirit."
5.—**Helen Almira**, b. April 4, 1812; m. Seagrove W. Magill, June 12, 1834; educated at a Young Ladies' School in N. Haven; res. Amherst, Mass. Rev. S. W. Magill, D. D., b. St. Mary's, Georgia, Sept. 27, 1810, entered Amherst Coll. 1827; grad. Yale, 1831, and studied theology, Princeton. Preached in Georgia, 1835-40; pastorate in Ohio and Vt. 1841-47; principal Female Sem., Athens, Ga., 1851; Cornwall parish, Vt., 1878; d. on his farm, Amherst, Mass., Jan. 20, 1884, of angina pectoris. *Issue:*
    William Alex., b. Jan. 2, 1836; m. Aug. 28, 1860, Matilda W. Smith; grad. Yale Coll. 1858; lives on his farm near Amherst, Mass.; four children.
6.—**Julia Webster**, b. Feb. 11, 1814.
7.—**Anne Loring**, b. Nov. 19, 1816; m. James Hadley, Aug. 13, 1851; b. M'ch 30, 1821, Fairfield, N. Y. He grad. Yale Coll., 1842; became Ass't Prof. of Greek, of Yale, 1848, and Prof. in full, 1851; d. Nov. 14, 1872. She res. N. Haven, Ct. *Issue:*
    One son, Arthur Twining, b. April 23, 1856; grad. Yale, 1876, and was appointed Prof. Political Science of Yale, 1876. A prominent writer on Railway Economy; res. N. Haven, Ct.

34 William Twining, son of **17** Thomas, b. Dec. 14, 1769, Eastham; d. Nov. 22, 1842, Tolland; m. Rebecca Brown, who d. Nov. 14. 1857, aged 82 years. Owned an extensive *farm in Tolland, Mass. Brought his bride of seventeen to their house on a pavillion the day they were m. Both lived and died in this house, wherein all their children were born and lived to manhood and womanhood. In 1810 was representative to the general court at Boston. Wife, Rebecca, was a sister of Col. Sanford Brown, who kept hotel at the foot of Tolland Mt.; family all Presbyterians.

*Mrs. Marcus Filley's account of her grandfather's (34 Wm.) homestead at Tolland, Mass.:

"'How in the name of wonder,' an expression of his, how he ever built a home in that wild, rough, out-of-the-way place, is more than I can imagine. Grandmother used to tell me that when they first lived there, bears and panthers would prowl around the house at night; the men were afraid to go out without their guns after dark. Their house was large—two or three kitchens with great, large fire places, (the children could sit in one corner). In the second story of the house was a large room for spinning and weaving; there was a large loom there. I think it is there now (1886); both wool and flax was manufactured in cloth for family use."

"G. F. had a large farm; raised horses and cattle. They kept about 30 or 40 cows; milking time was a busy one; every member of the family had a certain number of cows, every cow had a name, and the large dairy-room, with its churns and cheese presses, was a place of interest. He owned a saw mill, grist mill and dry goods store. In their large house, well filled as it was in my younger days, there was happiness; everything showed contentment and prosperity. But what a cold place in winter!—a house built upon a mountain. One could look off at the West five miles—cold and bleak, not a stove in the house. Every morning after breakfast (a fire built one morning would last until the next, if it did go out the tinder box was handy) the men would go out, and, with a crow-bar, roll in logs covered with snow; first would come a great back-log that would be rolled into its place; then another; then great chips on top of the logs; then the coals that had been left over would be put among the chips (all this time doors open, cold coming in), then the bellows were used until the fire was well started, and with all the exposure no one took cold. There was health and wealth on that mountain."

ISSUE:

1.—**Corinthia**, b. Oct. 9, 1793; d. M'ch 10, 1838, Otis, Mass. m. Lester Filley, Jan. 26, 1814. T.; moved to Otis, where he practiced law, and from there to Lee, Mass. Was a distinguished lawyer, member of State Senate, and filled other important offices. He established three Episcopal churches; m. secondly, Maria Wilcox, of Lanesboro, Mass. *Issue:*

   1. Caroline A., b. Feb. 6, 1815; m. Marcus L. Filley, lawyer; res. Lansingsburgh, N. Y.; Episcopal church.
   2. William Twining, b. Jan. 27, 1817; lawyer; Pittsfield, Mass.
   3. Hannah Roberts, b. June 13, 1822; d. 1840, L, N. Y.
   4. Henry Dwight, b. June 7, 1829; d. Oct. 1, 1852, Chester, Mass.; lawyer.
   5. Lester Bishop, b. Jan. 26, 1827, Otis; res. Troy, N. Y.; iron business.

2.—**Thomas**, b. Aug. 30, 1795; d. Nov. 14, 1865; Gt. Barrington, Mass. Grad. Williams Coll., Mass., 1814. Spent two years at Litchfield law school, and studied law with Sam. L. Jones, of Stockbridge, Mass., whose sister Rachel, a beautiful and accomplished woman, he m. fall of 1818; she d. 1850, dau. of *Gen. Jones of Hebron, Ct. Practiced law in Sandisfield to 1838. Was High Sheriff of Berkshire Co., Mass., many years; representative in Legislature 2 or 3 times. At coll. boarded in the family of Lyman Beecher. *Issue:*

   1. Lydia Rebecca, b. Aug. 5, 1819; d. Nov. 29, 1884; m. John W. Fibbits of New London, Ct.; chil. *Henry T., Katie W., Ralph G., Edward H.* Fanny C. and Harriett (Biggs) and one son d. inf.
   2. Rachel, b. 1821; d. inf.
   3. Samuel, b. April 19, 1822; d. Nov., 1843; coll. associate of Geo. P. Biggs, son of Gov. B.
   4. Clara M., b. Mch. 6, 1824; d. July 11, 1888; m. Leonard G. McDonald, a retired merchant, res. Glens Falls, N. Y. "She was a true, loyal, loving and noble wife and woman, whom scarcely one ever knew but to love and admire her noble character." Episcopal church, in which a very handsome memorial window is placed to her memory. No children.
   5. Thomas, Jr., b. Sept. 1, 1826; d. Sept. 6, 1826.
   6. Thomas A., b. Nov. 26, 1827; d. June 6, 1868; single.

*Gen. Jones was a comm. officer under George III. in the war of 1775, also under Washington in Col. Army, 1756, stationed at Lake George.

7. Emma B., b. June 18, 1832; m. Jno. Price, lawyer, who d. in Gt. Barrington, 1860. *Issue:*
    1. William T., b. Dec., 1854; ins. business, Hartford, Ct.
    2. Mary Alice, b. M'ch 27, 1859.
    3. Clara, b. Feb. 14, 1861.
8. Arthur H., b. June 5, 1836; d. Oct. 23, 1872, in Philadelphia hospital, result of wounds and army life in the war of '62-4; single.
9. —— d. inf.

3.—**Louisa,** b. June 17, 1797; d. June 6, 1866; m. Samuel Pickett, of Tolland. About 1838 moved from Otis, where he kept tavern, to Brooklyn, N. Y.; d. April, 1852. *Issue:*

    One daughter, Julia Louisa, b. Sept. 5, 1818; d. July 28, 1885; m. Hiram Sears, Sept. 26, 1838, Brooklyn, where he kept a wholesale boot and shoe store; moved to Vail, Iowa, 1879; one child, Lucia (Fitch).

4.—**Caroline,** b. Jan. 13, 1800; published to Samuel Cook, Sept. 25, 1820, Tolland. A well-to-do woman, whom everybody liked, and a "thorough Twining;" husband intemperate and left his family the last few years of his life; both d. in N. Y. City. *Issue:*
    1. Thomas, ——; m. a very nice girl, but soon after ran away.
    2. Abbie Ann, ——; m. Ray; she and her husband and little dau. were drowned when the steamer *Henry Clay* was burned on the Hudson river.
    3. Eveline, ——; m. —— Searle; three sons and two dau.; res. New York City.
    4. Sarah, ——; m.
    5. Chauncey, ——; m. and "disappeared;" whereabouts not known.
    6. Stephen, ——; d.
    7. Fredrick, ——; d.

5.—**Rebecca E.,** b. July 23, 1801; d. Aug. 27, 1850, Mich.; m. Chauncey Brown, Sept. 20, 1820; b. May 1, 1790, Homer, N. Y., lived in Avon, N. Y., until 1837, then moved to Genesee Co., Mich., until his death (in Flint), Aug. 5, 1864; farmer; Presbyterian. *Issue:*
    1. Hellen, b. Nov. 26, 1821, Avon; m. Rev. F. A. Blades, of Flint; M. E. Ch.; she d. Shiawasse Co., Mich., Sept. 8, 1849.
    2. Laura, b. July 14, 1825; m. in F., Jan 29, 1845, Francis King; b. April 30, 1820, Livingston Co., N. Y., son of James and Amanda; 1860 moved from Genesee Co., Mich., to Kent Co. (Lowell), present res. Lumber merchant; Cong.; three chil. and one niece adopted.
    3. Alexander T., and 4. d. in infancy.

5. Samuel P., b. May 31, 1831; d. M'ch 17, 1877, Marquette, L. S.; m. at Jackson, Mich., Nov. 18. 1857, Gertrude, dau. of William and Clarissa Wyckoff. Lived in Genesee Co, Mich., until 1868; architect and builder; M. E. Ch.; four chil.
6. Daniel, b. June 28, 1836; d. Boscobel, Wis., June 6, 1876; m. Oct. 6, 1859, Sarah Ritter, Green Castle, Ind.; 3 chil.
7. Chauncey C., b. Oct. 27, 1839, Grand Blanc, Mich.; d. May 27, 1862, in Lansing Agricultural Coll.

6.—**75 Alfred A.**, b. 1804; m. Marietta Hamilton.

7.—**Julia Ann**, b. Oct. 8, 1807; d. Sept. 19, 1872, Nunda, N. Y.; m. Feb. 5, 1827, by Rev. M. White, Jared Plum Dodge, Tolland; b. Aug. 25, 1800, Amsterdam, N. Y.; moved soon to Nunda, N. Y., present residence. Prominent business man and farmer; held many offices of trust; Presbyterian. *Issue:*

1. Alfred C., b. Feb. 14, 1830, Tuscarora, N. Y.; m. Katie E Bugen, Sept, 2, 1856; merchant; Nunda; 3 chil.
2. Chauncey B., b. Oct. 24, 1831; m. Sarah M. Kiggs, M'ch 12, 1856, Mt. Morris, N. Y.; d. of camp fever in army, July 6, 1862; his wife, first white child b. at Fenton, Mich., Oct. 12, 1836; Episcopal Ch.
3. William Twining, b. Sept. 9, 1834; d. Sept. 18, 1887, Willard Insane Asylum, N. Y., where he had been 4 mo.; m. Harriet Bugen, Nov. 25, 1858; b. Sept. 20, 1837, Scipio, N. Y.; chil.
4. Jared P., Jr., b. April 9, 1838, Union Corners, N. Y.; m. Julia Carpenter, Feb. 8, 1859; b. Aug. 22, 1839; chil.
5. Julia Louisa, b. M'ch 6, 1843, Tuscarora, N. Y.; m. Joseph Eastwood, April 5, 1865; b. May 10, 1839, Liverpool, England; resided Rochester, N. Y., St. Paul, Minn., and Bay City, Mich., 1866, present res.; merchant and lumber dealer; Episcopal church; one dau., Lucia L., b. Jan. 31, 1865, St. Paul.

The writer has had many delightful "pen talks" with "cousin Julia."

8.—**Sophronia**, b. Dec. 23, 1811; m. Aug. 12, 1835, Dennison Slocum, of Tolland; farmer, who d. Oct. 31, 1880, aged 81; wid. lives in T. (1889). *Issue:*

1. P. L——; res. T.
2. —— ——, a dau. who d.
3. Francis.
4. Joseph.
5. Alexander.

9.—**Stephen**, b. Oct. 12, 1812; m. *first*, Ann M. Hamilton (sister of his bro. **75** Alfred's first wife) about 1833, who d. Feb. 2, 1837, T., aged 22; m. *secondly*, Haphalonia Beach, M'ch 22, 1838, by Rev. William Harrison; b.

Feb. 14, 1814. In 1854, moved from Tolland to Tuscarora, N. Y., and to Nunda, 1857, present res.; farmer; d. 1888, Oct. 22d. *Issue* by 1st and 2d wives:
 1. Henry, b. Feb. 2, 1834; d. Sept., 1851, Tuscarora.
 2. Lycurgus, b. May 7, 1840; d. April 9, 1865; wounded at Beritonville and d. at Hospital, Goldsborough, N. C.; single.
 3. Corintha E., b. Feb. 14, 1842; d. Feb. 22, 1845.
 4. Romulus R., b. May 16, 1844; d. Feb. 29, 1845.
 5. Corintha E., b. July 15, 1847; m. Garrett Miller; grist mill, Tuscarora; one chil., Mary L., b. Aug. 3, 1875.
 6. Mary L., b. Jan. 7, 1853; d. Feb. 22, 1858.

10.—**76 Alexander H.**, b. Dec. 25, 1814; m. Laura Tinker.

35 William Twining, son of **18** Elijah; b. Nov. 13, 1763 E.: d. Nov. 12, 1846, Tolland; m. Tabiatha Smith, d. Jan. 25, 1854, aged 88 years. A very prudent woman. Farmer: buried on Tolland Mt.: chil. all born at Tolland.

ISSUE:

1.—**Betsey**, b. Sept. 13, 1787; m. Abraham Crane, who d. April, 1860; published Sept. 30, 1811; one son, Alexander, b. 1812, T.
2.—**77 William**, b. June 14, 1789; m. Ovanda Fowler.
3.—**78 Elijah**, b. Aug. 25, 1792; m. Almira More.
4.—**79 Hiram**, b. M'ch 31, 1794; m. Lovey Peace.
5.—**80 Joseph**, b. M'ch 27, 1796; m. Rachel Lewis.
6.—**Lucinda**, b. Nov. 9, 1798; d. 1885, Copenhagen, N. Y.; m. Levi Waters. *Issue:*
 Lyman, Joseph, Nelson, Robert and William; all living (1888), one in C.
7.—**Lyman**, b. April 5, 1801; d. Sept. 6, 1874, New Boston, Mass. A farmer, who held many town offices; became blind 5 or 6 years before death; Cong. Ch.; m. *first*, Paulina M. Shepard, b. June 30, 1805; d. July 1, 1833; m. *secondly*, Polly Henry, March 28, 1835; d. Feb. 20, 1875, aged 68 years; chil. only by first wife. *Issue:*
 1. Sarah, b. May 2, 1832; d. Aug. 22, 1864; grad. LeRoy (N. Y.) Female Acad., 1854; m. Seymour A. Tingier, of Webster, Mass., Nov. 25, 1857; b. at Tolland and d. July 23, 1888, East

Tompson, Ct., aged 58 years; grad. Williams Coll., 1855; members M. E. Church. Children: *Lyman Twining* (attorney), res. Rockville, Ct., and *Sarah Paulina*, res. Tolland, Mass.

    2. Paulina M., b. June, 1833; grad. Mt. Holyoke Female Sem.; d. Dec. 21, 1854.

8.—**Philina**, b. March 25, 1803; d. in Freedom, Portage Co., Ohio; m. W. Strickland; two daughters.

9.—**Nelson**, b. Dec. 25, 1806; d. Oct. 31, 1831; merchant, New Boston, Mass.

10.—**Milo**, b. 1812; d. Nov. 9, 1821, Tolland.

36 Eleazer Twining, son of 18 Elijah;

b. May 29, 1765 E.: d. May 30, 1829, Tolland: m. Mercy Smith, dau. of Eleazer of Sandisfield: d. Feb. 12, 1839. He was a devoted and influential member of the Con. Ch. At one time when dissentions had arisen in the church, he paid the pastor, Rev. Harrison's salary, amounting to two hundred dollars: farmer.

ISSUE:

1.—**Abigail**, b. Jan. 12, 1796; d. July 4, 1884; published to Loyal Humprey, Nov. 15, 1812. *Issue:*
    1. John, who d. in Va.
    2. Mary, m. James Daniels.
    3. Charles, m. four wives.

2.—**Ruth**, b. July 16, 1797; d. about 1878; m. Lester Moore, 1816; one child, Mary, who m. ——— Strawhacker.

3.—**81 Barnabas**, b. M'ch 1, 1800; m. Harriett Phelps.

4.—**Judah**, b. June 2, 1805; d. Sept. 1, 1885; m. Clarissa Smith, Sept. 7, 1838, Tolland; widow, res. Hartland, Mass. *Issue:*
    1. A daughter.
    2. Charles, d. April 20, 1863, from wounds received in the battle of Irish Bend.

5.—**Polly**, b. April 26, 1808; d. Sept. 17, 1882, Riverton, Ct; m. Lauren Smith, Feb. 22, 1831. Lost three children in youth. One son, Riley W., b. Jan. 22, 1836, m. Ann Cleveland, May 13, 1860.

37 Judah Twining, son of **18** Elijah, b. Jan. 21, 1774, E.; d. Aug. 27, 1854, Tolland; m. Catharine Fowler, Nov. 19, 1806; d. May 19, 1844, aged 67 yrs. A very benevolent man; devoted member of Cong. Ch.; noted for his fat cattle and hogs. Col. Samuel Pickett worked for him one summer. Always walked with two canes, owing to numerous corns upon the bottoms of his feet.

ISSUE.

1.—**Emoline**, b. Dec. 6, 1807; d. Dec. 20, 1807.
2.—**82 Philander F.**, b. May 6, 1809; m. Sarah A. Shepard.
3.—**Lois**, b. May 8, 1811; d. Mar. 16, 1836; m. Rev. Joel Talcott, Oct. 3, 1829; moved soon to Wellington, Ohio. Rev. T. b. Oct. 12, 1797, Vernon, Ct., and educated at Yale Coll.; 10 years minister at W.; d. Dec. 28, '71. *Issue:*
   1. Henrietta M., m. Joseph Twining, son of **78** Elijah Twining.
   2. Annete L., b. Wellington, April 19, 1834; m. *first*, John Foote, of Fitchville, Ohio, Oct. 13, 1857, who d. in Hartland, O., May 26, 1872; Baptist; one son; m. *secondly*, Henry Hammond, Oct. 13, 1873; res. Hartland, O.; farmers; Baptist.
   3. A son, who d. inf.
4.—**Cathrine**, b. Dec. 15, 1815; m. Dr. Samuel C. Parsons, Sept. 10, 1833. He practiced medicine 40 years in Western Mass.; went to Kansas City, Mo., 1873, and d. Oct. 3, '85, of paralysis, aged 76, wealthy. Wid. res. in K. City (1886). *Issue:*
   Hurburt C., Samuel I., Burton T., Cathrine J.; all dead but one grand child.

38 Lewis Twining, son of **18** Elijah; b. April 11, 1777, E.; d. July 18, 1821, Newark, Ohio; m. Jennett Smith, dau. of Deacon S., of Sandisfield, April 25, 1800; b. Jan. 4, 1780; d. Nov. 6, 1827, Delaware Co., Ohio. Sold the property his father gave him in north part of Tolland for six thousand dollars, and in 1815 moved to Granville, Ohio; invested in land and water privileges, which at his death was all lost through an avaricious creditor. The dam was afterward swept away and the land covered with gravel and dirt, so it was never after cultivated; supporter of Pres. Ch.

ISSUE: All but last b. in Tolland.

1.—**Almira**, b. Sept. 12, 1803; d. Dec. 20, 1883, Granville, Ohio; m. Rev. Samuel Rose, M'ch 31, 1826; b. July 25, 1800, East Granville, Mass. (His father, Timothy Rose, was one of the purchasing agents for and founder of a colony removing from Mass. to G., Ohio, in the fall of 1805; he was a judge and deacon of Pres. Church; d. 1813.) Entered Pres. ministry about 1828; grad. O. University; a prominent lecturer on theological and moral subjects; d. Jan. 10, 1857, New Lexington, O. His ancestry were Scotch and first settled at E. Granville, Mass., 1741. (While Abigail, his great grandmother, lived, she had over 450 living descendants; d. aged 103). *Issue:*

    1. Samuel L., b. May 28, 1827; d. at Chattanooga, of wounds received at the battle of Chicamauga, Oct. 21, 1863; Serg't 113 Ohio Vol. Inft., deacon of Pres. Church.
    2. Timothy D., b. Aug. 19, 1829; teacher and farmer. In 1862, Capt. Co. K., 96 Ill. Vol. Inft. In 1867, commission business, Chicago; 1871, Manistee, Mich.; 1887, res. Des Moines, Iowa; m. and has children.
    3. Almira J., b. Nov. 30, 1830; m. P. R. Eddy; res. Hartford, Ohio.
    4. Lydia, b. Feb. 18, 1833; res. Granville.
    5. Samantha M., b. July 20, 1840; m. Henry Hadley.
    6. Enoch A., b. July 3, 1846; d. July 4, 1879. Five other children died in infancy.

2.—**Lewis**, b. Aug. 14, 1805; d. July 4, 1821, Newark, O.
3.—**83 Merrick S.**, b. July 13, 1807; m. Corintha Clark.
4.—**Lauriston**, b. Nov. 9, 1809; d. Feb. 20, 1841, Fredonia, O.; m. Mary Robinson, M'ch 20, 1831; one son, who d. young.
5.—**84 Edward W.**, b. Oct. 5, 1814; m. Adelia Weed.
6.—**Darius N.**, b. Feb. 11, 1818; d. May 1, 1820.

39 Nathan Twining, son of **19** Jonathan; b. M'ch 8, 1755, Orleans, Mass; drowned at Alstead, N. H., at the age of 95; he went out in the evening near his house and fell into a brook, which resulted as above given; his early life was spent at sea; served throughout the Revolutionary struggle, after which he settled in N. H., where he married; removed to Alstead, N. H., prior to 1790, evidently not far from his first residence. Is said to have been "An honest man, much interested in religion and frequently took part in prayer meetings;" Cong.

"The Gilsum Hist., N. H.," states that he had four children; this, however, is disproved by his grandchildren, who claim he had only two; m. Sarah Clayton, who was found dead in bed by her husband at A.

ISSUE:

1.—**85** Jonathan, b. Nov. 9, 1790; m. Eliza Ann Fessenden.
2.—Tabitha, m. —— Clayton, her uncle. *Issue:*
   1. Sylvenus, who was blind.
   2. Richard, m. but had no chil.

40 Barnabas Twining, son of **19** Jonathan; b. May 14, 1767, Orleans; d. Jan. 3, 1847, in his 81st year at O.; m. Rebecca Rogers, M'ch 15, 1796; b. Nov. 17, 1769; d. Dec. 28, 1831. She was admitted to Orleans Church, July 31, 1803, at which time her three eldest children were baptized by Rev. Bascom.

ISSUE:

1.—Tabitha, b. Sept. 24, 1796, Orleans; d. Feb. 29, 1880, Boston; m. *first*, Joseph Cole, 1820; b. Eastham, Aug. 9,

1799; d. B., April 9, 1836; seaman; Baptist; m. *secondly*, Heman Crosby; b. Brewster, Mass.; d. Dedham, Mass., 1859; teamster, Unitarian. *Issue* by first m.

    1. **Rebecca R.**, b. Jan. 24, 1821; d. inf., O.
    2. **James T**, b. June 5, 1822; d. Oct. 19, 1831, B.
    3. **Joseph**, b. Aug. 20, 1825; d. July 29, 1826, B.
    4. **Joseph**, b. Oct. 9, 1826; d. Sept. 1, 1828, B.
    5. **George H.**, b. Aug. 30, 1829; d. Feb. 15, 1832.
    6. **James T.** (twin), b. Feb. 3, 1836; m. Roxannah F. Grandy, of Vt.; res. Boston Highlands, Mass.; children.
    7. **Joseph H.** (twin), b. Feb. 3, 1836; m. Sarah M. Eagles, of Novia Scotia; res. Jamacia Plain, Mass.; chil.

2.—**John**, b. Sept. 26, 1798; d. M'ch 22, 1850, Orleans almshouse; unm.; weak intellect.

3.—**86 Ebenezer**, b. April 4, 1801; m. Meribah Small.

4.—**James** (twin), b. July 11, 1804; d. June 30, 1805.

5.—**Joel** (twin), b. July 11, 1804; d. Aug. 12, 1852, Orleans almshouse; unm.; weak intellect.

41 Abner Twining, son of 20 Barnabas;

b. Jan. 20, 1772, Eastham; d. Feb. 1, 1850, Frankfort, Maine; m. *first*, Mary, dau. of Heman Snow, Dec. 19, 1793; b. Nov. 5, 1769, E.; d. June 18, 1851; m. *secondly*, Mrs Small, Nov. 8, 1851. "He was a remarkable man, above the common stamp in regard to intellect and acquirements." He could always "carry the day" at Town and other meetings. Moved soon after m. to Frankfort, Maine, where he was engaged in teaching in the public schools for many years, and filling, at different times, town offices. Was wealthy, owning an extensive farm cleared in the wilderness; he and his descendants about all Universalists.

ISSUE:

    1.—**Mary**, b. Oct. 29, 1794, F.; m. Edward Snow about 1819, F. They lived and d. at F.: farmer. *Issue:*

THE TWINING FAMILY.

 1. Williamson Twining, b. June 1, 1820; d. Brookfield, Mo., June 26, 1886, where he was P. M. 20 years. Moved to Mo. 1858; merchant; children.
 2. George, b. about 1824; d. 1878, Bangor, Maine; Capt. of a steamboat for many years; a man of honorable mention; children.
 3. Elvira W., b. about 1826; m. and kept hotel 35 years in Union, Me.; a very estimable woman; res. U.
 4. Henry O., b. Jan., 1830; m. and has one of the finest farms in Waldo Co , Me.; res. Winterfort, Me.; 4 chil.
 5. James; a well-known citizen of Brewer, Me.; one chil.
 6. Albert, d. aged about 21 yrs.

2.—**Abigail**, b. Nov. 25, 1796, F.; m. Jeremiah Wardwell. Both d. at Winterport, where an only son, James Wardwell, a prominent citizen, res.

3.—**Thankful**, b. July 28, 1798; d. Oct. 1, 1848, unm.

4.—**Chloe**, b. Feb. 25, 1800; m. James Downs. A dau., Mrs. R. A. Shaw, is the wife of one of Winterfort's most prosperous farmers.

5.—**Abner, J.**, b. Nov. 6, 1801; d. Nov. 16, 1849: single.

6.—**David**, b. April 4, 1803; d. M'ch 31, 1848: single.

7.—**Nathan**, b. June 12, 1805; d. about 1867; m. Lucretia Woodman, of F., Dec. 4, 1834; d. about 1877; farmer near F.: no issue.

8.—**Sarah**, b. Sept. 5, 1808; d. Dec. 1, 1826: unm.

9.—**87 Addison**, b. June 24, 1810; m. Emeline Colson.

10.—**Williamson**, b. May 25, 1813; d. M'ch 5, 1816.

11.—**88 Harrison**, b. Nov. 14, 1814; m. Olive Higgins.

42 Jonathan Twining, son of **21** Prince; b. M'ch 25, 1775, Orleans; d. 1799 "in the war or at sea," where he went soon after m.; m. Tamzin Snow, M'ch 2, 1797, O. In 1800 she went to Pig Island Maine, where she m. Mr. Lake about 1809, then moved to Powell, Me.

     ISSUE:

1.—**89 Jonathan**, b. May 13, 1799, O.; m. Sukey Linnell.

2.—**Tamzin** (twin), b. May 13, 1799, O.; went with her mother to Maine when nine months old; m. Abner Knight, of

Powell, Me., and removed to East Boston, Mass., about 1840, where she d. 1883: he d. some years previous. *Issue:* Two children.

43 Prince Twining, son of **21** Prince; b. April 30, 1783, O.; d. Dec. 25, 1860, Lisle, Broome Co., N. Y.; m. Mary, dau. of Capt. Seth Higgins, Feb. 18, 1811. A sailor up to 1818, when he moved from Orleans to Boylstown, N. Y. About 1820 settled in Broome Co., where he was engaged in farming remainder of life. Was an intelligent man, well proportioned, medium size. She was "aristocratic in feeling;" d. after 1860: Presbyterians.

ISSUE:

1.—**Seth H.**, b. Feb. 20, 1812, O.; m. *first*, wid. Parnell Fox, Feb. 20, 1836; d. in Oneida Co., N. Y., Oct. 8, 1862; m. *second*, Sarah Eddy, "an old school teacher and a splendid writer," Jan. 17, 1880. Removed from Broome Co. to Oneida Co., N. Y., about 1860; farmer; at one time kept hotel at Lenox. A heavy set man, broad shouldered, athletic, and above the average in business ability. These were characteristics of his brothers; Free Thinker; blind in latter life; res. Oneida, N. Y. *Issue* by first wife:
    1. Seth R , b. July 29, 1841; d. Jan. 5, 1863, in Washington, D. D., hospital during army service; member Co. H., 117th Reg. N. Y.
    2. Lyman E., b. May 3, 1843; d. Nov. 11, 1858.
    3. Ellen J., b. Oct. 23, 1846; d. Nov. 16, 1858.
    4 Emma S , b. May 1, 1848; d. Oct. 3, 1862. *Issue:* by second wife:
    5. Florence Eda, b. April 20, 1881.
2.—**Francis**, b. Feb., 1814, O.; d. Sept. 28, 1870, Auburn, N. Y.; m. Sarah J. Dakin (wid.) Feb. 8, 1839, Geneva, N. Y.; d. Sept. 18, 1870. Run a line of stages from G. to Lyons at one time. In 1844 to 1867 kept hotel, Cayuga, N. Y. A man of good character, quiet and unobstrusive. *Issue:*

THE TWINING FAMILY.   81

1. Frances S., b. Nov. 17, 1839; m. M. D. Townsend, of San Francisco, Cal., M'ch 12, 1861; d. June 11, 1880; a speculator in mining stocks; children; wid. res. San Leandro, Almeda Co., Cal., formerly of Coffin's Summit, N. Y.
2. Josephine A., b. Aug. 28, 1841; m. John A. Davis, of Springport, N. Y., June 11, 1867; chil.; res. Springport, N. Y.; farmer.
3. Edgar O., b. July 8, 1844; d Oct. 27, 1879, of typhoid fever, Michigan City, Ind.; in the army, but taken prisoner at Harper's Ferry and released at Chicago, Ill.; unmarried.

3.—**Joseph B.**, b. M'ch, 1816, O.; d. June 17, 1854, Lisle, Broome Co., N. Y.; m. Eliza Stoddard, Sept. 9, 1841, by Rev. Babbitt, dau. of James and Rhoda (Hyde) Stoddard. Followed lumbering several years, and then kept hotel at Binghamton, N. Y. Wid. again, m. James Rowland, and about 1866 moved to Ill. Is now (1888) wid. again; res. with her eldest dau.; res. Minneapolis, Minn. *Issue:*
  1. Ellen, m. Philo Meachem; res. M. Minn.
  2. Alice.

4.—**Mary**, b. doubtless at Orleans; d. 1859, Lisle, N. Y.; m. *first*, probably —— Brown, and *secondly*, Geo. Owen, who was a Lieutenant in 137th Vol. N. Y., and shot at the battle of Lookout Mt. *No issue.*

5.—**Lucy**, b. 1821; m. *first*, Lathrop Rood (deceased); m. *secondly*, —— Gaylor; farmer; Presbyterian. "A fat, fussy, feeble-minded old woman;" res. Centre Lisle, N. Y. *Issue:* by first husband:
  1. Andrew, d. July, 1878, Cortland, N. N.; m. Miss Roop (cousin). Children: Ina, Hattie and Florence; res. C.
  2. Reuben; d. Centre Lisle.
  3. Frank; d. Centre Lise.
  4. Anna, b. about 1854; m. —— Lusk. Chil.: Fred. and Ed.
  5. Mary, d.

6.—**Jane**, d. Oneida Co., N. Y., at the birth of a son; m. Warren Potter. One child, Jay Potter, who res. Stockbridge, N. Y.

7.—**John**, d. inf.

SEVENTH GENERATION.

44 Charles Twining, son of **22** Stephen; b. Troy, N. Y., Aug. 9, 1820; m. *first*, Eliz. H. West, at Forrest Particular Meeting, Dec. 7, 1842, dau. of Mahlon and Mary, of Harford Co., Md.; b. Jan. 29, 1821; d. Yardly, Bucks Co., Feb 3, 1884; m. *secondly*, Hannah Y. Bunting, Sept., 1886; b. Nov. 8, 1823, dau. of Joseph and Sarah (Yardley) Paul; Friends; "burner and shipper of lime;" res. Newtown, Pa.

ISSUE.

1.—**90** Stephen B., b. Jan. 19, 1844; m. Latetia Warner.
2.—**91** Edward W., b. M'ch 4, 1846; m. Mary Smith Walker.
3.—Mary E., b. M'ch 27, 1849; m. Franklin Eastburn, Oct. 28, 1869; farmer; res. Edgewood, Pa. *Issue:*
    1. Sarah C., b. M'ch 26, 1871.
    2. Charles Twining, b. Sept. 10, 1873.
4.—Emma, b. Aug. 25, 1851; m. R. Franklin Schofield, June 1, '76; farmer; res. N. *Issue:*
    1. William L., b. Feb. 28, 1881.
    2. Edw. Twining (twin), b. Feb. 7, 1884.
    3. Joseph (twin), b. Feb. 7, 1884.
5.—Charles P., b. Jan. 13, 1854; d. Jan. 21, 1864.
6.—Rebecca R., b. M'ch 7, 1856; m. Watson G. Large, Feb. 10, 1886; res. Yardley, Pa. *Issue:*
    1. Eliz. Twining, b. M'ch 14, 1887.
7.—William R., b. April 4, 1858; d. Jan. 25, 1864.
8.—Adeline, b. June 14, 1860; d. Oct. 25, 1862.
9.—Sarah S., b. Sept. 26, 1862; d. M'ch 19, 1864.

45 John Twining, son of **23** Thomas; b. Dec. 2, 1784, Quakertown, N. J.; d. Waterloo, Jefferson Co. Wis., Feb. 8. 1875; m. *first*, Sarah Palmer, Feb. 15, 1807, Q.; b. Dec. 10, 1787, Q.; d. Boston, Erie Co., N. Y., M'ch 4, 1825; m. *secondly*, Sarah, dau. of *Aaron and Jane Hampton, June 7, 1826, Boston; d. Feb. 5, 1883, Waterloo; b. Aug. 4, 1807.

" His was an earnest, toiling, struggling life, busy and full of good works. Overlaping the fourth generation from his birth, accumulated experiences had given him wisdom rarely excelled. The honored recipient of a public office for fourteen successive years, he grew in public favor and trust. A volunteer of the 1812 war; a pioneer in western N. Y., he felled the forest and reared with his own hands shelter for his increasing family. His life four times in jeopardy, once from fire, twice from the assassin, and oft in battle, he was graciously spared to rear his family and bless his friends. Powerful in debate, eloquent in conversation, an inveterate hater of slavery, he exerted all his influence to its downfall and extinction."

Came with his father to western N. Y., where he followed farming until Aug., 1844, when he settled near Waterloo, Wis., where he lived the remainder of his life. Owned much land. A remarkably strong and healthy man through life: Quaker.

ISSUE BY FIRST WIFE:

1.—**David**, b. Sept. 9, 1807, Quakertown, N. Y.; d. Aug. 1865, Lima, Rock Co., Wis.; farmer: single.

---

* Aaron Hampton, b. 1780, N. J.; moved to Boston, Erie Co., N. Y., about 1816, where he d. 1855. A "Quaker of the most faithful kind;" farmer, wheelwright and nurseryman. His father, David, b. 1757, Bucks Co., Pa., and d. there 1790, son of John and Ann (Croasdale) Hampton, b. 1724; d. 1775. She was a minister of Friends Society; d. 1796, all of Bucks Co., Pa. John Hampton was the son of Joseph and Mary (Camby), of N. J., who came to Bucks; m. 1722, where he d., 1794, aged 97.

2.—**Charles**, b. Feb. 20, 1809; d. 1881, Buffalo, N. Y., unsound mind: lived with Levi Bunter, Eden, N. Y.: single.

3.—**Susan**, b. Nov. 20, 1810, Willink, Erie Co., N. Y.; d. Feb. 20, 1851, Wis.; m. John Webb, 1833. A kind and gentle woman. *Issue:*

    1. Clark T., b. Jan. 27, 1834; m. Nov. 1858, Harriett E. Lancaster; res. Union Centre, Wis.; chil.
    2. John; d. infancy.
    3. Geo. W., b. Feb. 1, 1838; d. M'ch 2, 1878; m. Nellie Norcoss; res. Fort Atkinson, Wis.; merchant; Cong.
    4. Horace; d. young.
    5. Henry, b. May 18, 1842; d. Sept. 26, 1862.
    6. John; d. young.
    7. Alvira, b. Nov. 3, 1846; m. John Hoag, Sept. 15, 1869.

4.—**Thomas**, b. June 12, 1812; d. June 18, 1815.

5.—**Sarah A.** (Sally), b. Feb. 10, 1814; d. Feb. 3, 1838; Eden, Erie Co., N. Y.

6.—**Thomas**, b. May 25, 1816; d. M'ch 20, 1864, Medina, Wis.; m. his brother, John Heely's, widow, Oct. 27, 1846; farmer; converted on his death bed. *Issue:*

    1. Ellen Sophia, b. Dec. 15, 1847, Waterloo, Wis.; m. Samuel R Seaver, Nov. 11, 1866; b. Oct. 27, 1834; druggist; res. Tecumseh, Neb. Children:

        1. Arthur Twining, b. Sept. 5, 1867.
        2. Edwin H., b. Sept. 22, 1869.
        3. Burt E., b. Jan. 20, 1872.
        4. Dora E., b. Feb. 18, 1874.
        5. Sarah E., b. Sept. 1, 1877.
        6. Samuel R., b. April 11, 1884.

    2. Alice Arvilla, b. Nov. 24, 1853; m. Sam. Stoffer or Spoffer, 1878; chil.: Don Lewis, who d. young and Victor, b. Feb. 27, 1881.

7.—**John Heeley**, b. Feb. 4, 1818; d. July 27, 1845, Milton, Wis.; m. Ann, dau. of Capt. Lewis, of Tubbs Hollow, Erie Co., N. Y., Dec. 25, 1838. "He acquired a good education, was a ready talker, a good debater, a good civil engineer and the People's candidate for Territorial Legislature of Wis. at the time of his death. Powerful in body, robust in health, he was stricken down in a day." His wid., Ann, b. July 4, 1817; d. Jan. 22, 1868; m. Thomas Twining (brother) Oct. 27, 1846; M. E. Ch. *Issue:*

1. Susan A., b. April 9, 1840, Boston, Erie Co., N. Y.; m. *first*, Theo. Weed, Dec. 30, 1854, Waterloo, who d. Nov. 24, 1864, at Cairo, Ill., while on his way home from the army; m. *secondly*, Wm. R. Roach, Sept. 16, 1866; native of Penn.; farmer; res. Menominee, Wis. Children by first husband:
    1. Geo. H., b. May 7, 1856, Adams Co., Wis.
    2. Nellie A., b. July 12, 1858, Waterloo; d. July 30, 1859, Kan.
    3. Adelia, b. May 5, 1861, Dane Co., Wis. By second husband:
    4. Mary A., b. Dec. 12, 1867, W.
    5. Hettie, b. M'ch 12, 1872, W.
    6. James W., b. Oct. 30, 1874, W.
    7. Frank, b. July 8, 1877, Hay River, Wis.
    8. Florence M., b. July 6, 1881, Sherman, Wis.
2. John Quincy, b. June 26, 1843; d. Pilot Knob, Mo., M'ch 22, 1862; member of Co. C, 11th Wis. Inf.; unmarried.

8.—**Margaret**, b. M'ch 14, 1820; d. M'ch 9, 1825.

9.—**Jane**, b. Sept. 4, 1822; m. *first*, Simeon Griffith, 1840; b. Aug. 1821; d. Jan. 14, 1861, Medina, Wis.; farmer; M. E. Ch.; m. *secondly*, Munson Tousley, M'ch 24, 1862; b. M'ch 16, 1814; druggist; d. Feb. 4, 1889; res. Olivet, Wis. She is member of M. E. Ch. and he of the "Christian Science Healer." *Issue:*
1. Irwin, b. about 1841.
2. Eleanor, m. William Bowers; res. Waterloo, Wis.; chil.: Ida J., Julia E., Chloe E., Amos F., Alice, James and Lewis.
3. William, d. inf.
4. Ann M., d. inf.
5. Julia A., b. Feb. 1, 1850, Stony Brook, Wis.; m. twice.
6. Albian, b. June 3, 1863; m., res. Warren, Ill.

10.—**Marvin**, b. Aug. 25, 1824; drowned, July, 1844, Rock Co., Wis.; unm.

ISSUE by second wife, b. in Erie Co., N. Y., except last one:

11.—**92 Aaron**, b. M'ch 31, 1827; m. Mary Lyons.

12.—**James**, b. Sept. 23, 1828; d. June 4, 1829.

13.—**Hugh**, b. Feb. 26, 1830; m. Nov. 29, 1867, Almira A. Fields; b. Aug. 8, 1841. Rambled about in Nev. and Cal. prior to m.; merchant; Quaker; res. Georgetown, Col. *Issue:*
1. Sarah L., b. M'ch 24, 1869, Medina, Wis.
2. Florence A., b. Nov. 15, 1873, Medina, Wis.
3. Warren H., b. Jan. 12, 1876, Medina, Wis.

14.—**Elizabeth**, b. June 3, 1832; m. Calvin Perry, Nov. 22, 1848; b. Nov. 3, 1827; farmer; M. E. Ch.; res. Fort Atkinson, Wis. *Issue:*

    1. Jennie, b. Dec. 2, 1853, Dodge Co., Wis.; m. Jan. 20, 1874, Geo. Stephenson; b. Nov. 13, 1845; farmer; res. Fort Atkinson.

    2. Harrison E., b. Sept. 29, 1864, Adams Co., Wis.; m. June 20, 1886, Mary A. Lieberman, b. May 3, 1866; teacher; preparing for the ministry; M. E. Ch.; res. Fort Atkinson.

15.—**93 Nathan C.**, b. Sept. 27, 1834; m. Phebe Ann Barber.

PROF. N. C. TWINING, SR.,
Supt. and Prin. Riverside Public Schools, Cal. Maj. 3rd Cal. Reg't.
(See under family head, 93.)

16.—**Margaret**, b. July 11, 1836; d. April 30, 1838.

17.—**Phineus Elijah**, b. Feb. 7, 1839; m. Feb. 1864, Jane E. Thomas. (See **46** Charles.) Enlisted same month (Feb.) in 36th Wis. Vol., 1st Serg't Co. F., and was wounded in the battle of the Wilderness, June 3, '64, and d. Oct. 15th following at Philadelphia, Pa., his brother, Nathan C., being with him when he d. and brought his body home.

His last words were: "I have given myself a willing sacrifice. I die content. If the enemy is subdued, the Union restored, you need not regret my death. Bury me in the home cemetery." "Was liberally educated, of pleasing address and commanded the highest respect of all who knew him. A very successful school teacher;" Baptist; home, Waterloo; no children. "Requiescat in pace."

18.—**94 Henry Harrison**, b. Feb. 11, 1841; m. Hattie E. Miller.
19.—**95 Peter Slater**, b. Feb. 27, 1844; m. Cornelia Z. Cooper.
20.—**John, Jr.**, b. Feb. 20, 1846, Waterloo; d. Jan. 5, 1854.

46 Charles Twining, son of **23** Thomas; b. July 20, 1789, Quakertown, Pa.; d. Farmington, Warren Co., Pa., April 18, 1871; m. Betsey Boutwell about 1811; b. July 4, 1785; d. Sept. 6, 1871, F. Moved from Erie Co , N. Y., where his children were born, to Potoci, Grant Co , Wis., about 1855, where he resided on his farm to within a few weeks of his death, which occurred as above stated, from bleeding at the lungs. Was a cripple from boyhood, caused by an injury to a foot by a horse, and always used crutches; tailor by trade; Quaker. None of his family could read or write.

ISSUE:

1.—Amanda, b. May 19, 1812; m. Thomas Widdifield, b. Nov. 22, 1811; d. Nov. 28, 1887; farmer; Quaker; res. Russell, Warren Co., Pa. *Issue:*
    1. Charles, b. Aug. 5, 1834; m. ——; res. Russell, Pa.
    2. Lydia A., b. M'ch 30, 1836; d. Sept. 27, 1889; m. Wm. Way; res. R.
    3. Thomas J., b. June 29, 1838; farmer; m. twice.
    4. Sarah J., b. M'ch 2, 1840; d. Sept. 18, 1858.
2.—**Betsey E.**, b. 1814; m. Sept. 27, 1834, John Hampton, brother to the wife of **45** John Twining; d. Oct. 14, 1870; she d. 1873, Wis. *Issue:*

1. Jane E., b. July 2, 1835; m. *first*, Theo. Thomas, who was drowned in Wis. river; m. *secondly*, Elijah Twining, son of **45** John, Feb. 23, 1864; m. *thirdly*, Fulton L. Miner, M'ch 7, 1866; res. Cedar City, Mo.; 6 or 7 chil.

2. Sophia, b. Nov. 19, 1837; m. Phineas Walker, Nov. 26, 1857; res. Lancaster, Wis.; chil.

3.—**Thomas**, b. Dec. 16, 1815; m. *first*, Udora Walker about 1842, Erie Co., N. Y. They separated and he married again, Jane Morgan, and settled in Warren Co., Pa.; res. Russell; farmer. *Issue* by first wife:

    1. Thomas Q., d. Helena, Ark., Aug. 14, 1863; member of 5th Wis., V. I.

    2. Ellen, lives with her mother in Hamburgh, Erie Co., N. Y.

4.—**96 Chapin**, b. Jan. 16, 1817; m. wid. Kee.

5.—**Elwood**, b. June 29, 1819; returned from Cal. about 1866 to Pa., blind and has been in Warren Co. asylum ever since; unm.

6.—**John**, b. Sept. 15, 1821; d. Aug. 29, 1862, serving in the Federal army at Arlington Heights, Va.

7.—**Mary S.**, b. Dec. 11, 1823; m. Amandus Sherwood, April, 1856; resided at Marshall, Dane Co., Wis., where they both died, 1857; no chil.

47 Thomas Twining, son of **23** Thomas; b. Jan. 13, 1801, Quakertown, N J.; m Sarah, dau. of *Ben. and Rachel Kester, Sept. 27, 1820; b. Dec. 19, 1799; d. near Huntington, Ind., Dec. 5, 1843

*Ben. Kester, who m. Rachel, dau. of Stephen and Hannah Hambleton, was a son of Samuel Kester, b. Sept. 26, 1737; d. M'ch 8, 1804, and of Susan Webster, his wife, b. Oct. 11, 1736; d. Feb. 24, 1832.

Of Samuel Kester, we learn from the *Hambleton Genealogy* that one dau., Sarah, m. Joseph Palmer, of Chester Co., Pa., and a dau., Susannah, m. Joseph Stevenson; lived at Rahway, N. J., and d. there a few years since in her 99th year.

He was of German descent, lived near Quakertown, N. J., and he and his family were prominent members of that Monthly Meeting.

"Skinner Webster, who m. Jane, a sister of Rachel above mentioned, and several Websters, who moved to Erie Co, N. Y., and there intermarried with Hambletons, were closely related to Susannah Webster above named."

Came with his parents to North Boston (Podunk), Erie Co., N. Y., 1811. At this place he m., reared his family, farmed, kept store, practiced medicine (Tompsonian system), etc. Moved with his family to Huntington, Ind., on the Wabash, where he bought 200 acres of choice land in the "forks" of the Wabash and Little rivers, a famous resort and place of council for Red men. Started by overland Sept. 20, '41, arriving there Oct. 9th following.

The land at this date is valuable, being underlaid with the best of limestone; the scenery is picturesque. At that date H. was a mere hamlet of log huts, Indians numerous, and wild game abounded. In 1890 the town has become a place of 8,000, extending to the "old homestead." Because of the dreaded "fever and ague" he became sick of the country, and in 1846-7 returned to N. Y., where he has been engaged in the dairy business on his farm in North Collins Tp., Erie Co., N. Y.; P. O. Lawton Station. Though just passing to his 90th milestone, his health is good for one of extreme age.

Became lame in the knee in his younger days and remained so during life. A slender man, above the medium height, active, well informed, ready conversationalist and much devoted to Friend's principles (orthodox), which he occasionally preached in latter life. Was also custodian of meeting records.

The compiler of these records has frequently seen grandfather ride to his first and fourth day meetings, some three miles away, through the storm and bitter cold to worship with the "two or three" equally loyal ones who had gathered to uphold the cherished tenets of Quakerism.

AUTOGRAPH OF **47** *Thos Twining*

Ben. Kester, b. in Kingwood, N. J., and d. 1819. In 1812 moved to Boston, Erie Co., N. Y.; Quaker and farmer; m. Rachel Hambleton, June 19, 1782; b. M'ch 7, 1765; d. M'ch 6, 1858, aged 93 years. Her grandfather, *James Hambleton*, who d. in Solebury, Bucks Co., Pa., July, 1751, is the first known ancestor of the Bucks Co. Hambletons. (See Hambleton Gen., 1887, Chicago, Ill.)

THE TWINING FAMILY.

ISSUE:

1.—**Melissa**, b. June 22, 1821; d. July 3, 1846, Huntington, Ind.; m. Joel P. Seeley, Sept. 6, 1843; b. Aug. 27, 1819; d. Sept. 1, 1889, of consumption; miller by occupation. *Issue:*
> One child, Thomas, b. July 3, 1846; d. infant.

2.—**Rachel**, b. Feb. 9, 1823; m. Joel P. Seeley (above), Nov. 30, 1848; Baptist; res. East Pembroke, Genesee Co., N. Y. *Issue:*
> 1. Melissa A., M'ch 13, 1850; m. John Allen, Dec. 2, 1868; b. Dec. 2, 1842; two children; res. E. Pembroke, N. Y.
> 2. Wilgus E., b. June 4, 1852; d. inf.
> 3. Charles H., b. May 28, 1854; m. Blanche Carpenter, Jan. 27, '81, N. Y.; b. 1856; one dau.; real estate and loan; res. Faulkton, S. D., where he owns 300 acres of land.
> 4. Clara A., July 17, 1856; m. June, 1884, Wm. A. Davis: res. Chattanooga, Tenn.
> 5. Emma M., b. July 13, 1858; d. inf.
> 6. Ida V., b. June 4, 1860; d. Dec. 26, 1870.
> 7. Arthur J., b. July 7, 1865; printer; res. Detroit, Mich.

3.—**97 Dewitt C.**, b. Sept. 23, 1824; m. Susannah G. Hambleton.

4.—**Mary A.**, b. Nov. 17, 1826; d. Aug. 14, 1827.

5.—**Selinda**, b. Feb. 23, 1828; m. *first*, Marmaduke Battey, Feb. 23, 1845, Huntington, Ind.; b. Sept. 9, 1820, E. Hamburg, N. Y.; d. M'ch 20, 1849, Lagro, Ind.; daguerreotypist; m. *secondly*, Paris Sprague, April 13, 1859; b. May 19, 1816, Vt.; d. Aug. 10, 1867; farmer; res. Glenwood, N. Y.; Free Will Baptist. *Issue* by first husband:
> 1. Sarah M., b. Oct. 25, 1847, N. Boston, N. Y.; m. Joshua Churchill, Dec. 14, 1864; farmer; Baptist; children; res. Boston Centre, Erie Co., N. Y.
> 2. Estella V., b. Dec. 18, 1849, Lagro, Ind.; m. April 10, 1867, Norman Freeman; miller; res. Randolph, N. Y.; 3 children. *Issue* by second husband:
> 3. Earl E., b. April 5, 1860, Concord, Erie Co., N. Y.
> 4. Arthur E., b. May 28, 1862; res. Glenwood.
> 5. Kittie B., b. July 17, 1867; res. Glenwood.

6.—**98 Lewis**, b. M'ch 24, 1830; m. Mary E. Sherman.

7.—**Anna**, b. M'ch 20, 1837; m. Joseph Manchester, Jan. 28, 1857; b. Jan. 8, 1837; farmer; res. Concord, Erie Co., N. Y. *Issue:*

1. Elmer E., b. Aug. 3, 1860; m. about 1880, Loittia A. Foster; res. Fountain, Mason Co., Mich.; one child, d. 1886.
2. Delbert D., b. Aug. 5, 1863.
3. Norman V., b. Jan. 28, 1867.
4. Lincoln G., b. Nov. 9, 1871.
5. Bertha B., b. Jan. 21, 1875.

8.—**Maryette**, b. Sept. 28, 1838; m. Orlando Luther, farmer; res. near Eden, Erie Co., N. Y.; one child, Luella, who m. —— Smith and res. near Eden; children.

48 Thomas Twining, son of **24** John; b. Sept. 4, 1790, Hunterdon Co., N. J.; d. Groton, Tompkins Co., N. Y., 1863. Came with his uncle, **23** Thomas, to Erie Co., N. Y., and worked for him about one year; then enlisted in the 1812 war, with headquarters in Tompkins Co.; m. Elizabeth McKenzie about the close of the war; she d. June 13, 1843, aged 53 years.

ISSUE:

1.—**John Ferdinand**, b. about 1816, Groton; d. in the West, 1847: unmarried.
2.—**William**, b. July 5, 1818, G.; m. Almira A. ——, who d. April 9, 1887, aged 64 yrs. A local M. E. preacher until about 1872, when he identified with the U. B. Church; also a farmer; res. Clymer, Chautauqua Co., N. Y.; no issue.
3.—**Artemas**, b. Sept. 20, 1820, G.; d. Nov. 1865; m. Phebe Cotanche, Nov. 5, 1850: b. Feb. 12, 1829: res. King's Ferry, Cayuga Co., N. Y. *Issue:*
    1. Charles, b. Oct. 7, 1851; d. Oct. 1866, Wisconsin.
    2. Augusta C, b. June 19, 1854; m. Jno. N. Starner, M'ch 18, 1874; life ins. business; res. Auburn, N. Y.
    3. Nancy A., b. July 27, 1856; m. Newell G. Coon, Jan. 1884; res. Ledyard, Cayuga Co., N. Y.
    4. Mary E., b. May, 1858; d. 1860.
4.—**99 Charles**, b. Aug. 23, 1822: m. Mary Stanton.
5.—**Olive**, b. Sept. 8, 1826, G.; m. Simeon Castle, 1847; b. 1823; res. Syracuse, N. Y. *Issue:*
    1. Charles, b. 1848.
    2. Alice, b. 1853.

6.—**Prudence A.**, b. Oct. 10, 1831, G.; m. Daniel D. Dimon, Jan. 2, 1853; b. 1829; d. 1884; member G. A. R.; res. Groton Village, Tompkins Co., N. Y. *Issue:*
    1. Wm. Frank, b. Feb. 9, 1854; m. Helen Waule, July 4, 1875; farmer; res. G.
    2. Fred., b. April 7, 1856; m. Anne McGregar, Dec. 19, 1883; farmer; res. G.
    3. Willie, b. June 21, 1858; m. Kate Waule, Feb. 8, 1885; farmer; res. G.

49 John Twining, son of **24** John; b. M'ch 25, 1794, N. J.; d. Broome Co., N. Y., Jan. 21, 1867; m. Dorcas Fonner, b. Jan. 20, 1792; d. Jan. 25, 1867. Rafted logs and lumber down the Susquehanna river before railroad times; adm. of his father's estate.

ISSUE:

1.—**Rachel**, b. July 17, 1815; m. Nov. 15, 1832, —— Davis; d. M'ch 22, 1885. *Issue:* John, Alvin, Lucy, Matilda, Dorcas and Eliza.
2.—**Leah**, b. Jan. 15, 1816; m. Oct. 23, 1834, —— Lebanon; d. Jan. 26, 1879. Children: *Asel*, Alvin and Alvira, all of Union, Broome Co., N. Y.
3.—**100 James**, b. Aug. 10, 1817; m. Rebecca Howard.
4.—**101 Thomas**, b. Aug. 11, 1819; m. Lucy Balch.
5.—**102 William**, b. Sept. 23, 1822; m. P. R. Miner.
6.—**Mary Ann**, b. Jan. 23, 1823; m. —— Choat; res. Seneca, N.Y. *Issue:* Charles, Josephine, William, Howard, Lyle, Mary and Seward.
7.—**Eliza**, b. Jan. 20, 1825; d. Feb. 23, 1850; unm.
8.—**103 John A.**, b. Oct. 16, 1827; m. Emily Roberts.
9.—**104 Charles**, b. April 16, 1831; m. Lucy A. Gibbs.
10.—**105 Philip**, b. Aug. 13, 1833; m. Francis A. Councilman.
11.—**Dorcas**, b. May 14, 1836; d. Nov. 2, 1839.
12.—**Rebecca**, b. Oct. 3, 1838; living (1885); unm.

50 Samuel Twining, son of **24** John, b. Feb. 22, 1796, Hunterdon Co., N. J.; d. April 10, 1831,

Broome Co., N. Y., where he was engaged as farmer and miller; m. Sept. 23, 1815, Eliz. Stout, b. July 7, 1797; d. Oct. 17, 1882.

ISSUE:

1.—**106 Joseph N.**, b. Nov. 12, 1818; m. Ruth A. Ames.
2.—**107 Charles A.**, b. May 23, 1821; m. Nellie Schermerhorn.
3.—**Rozette**, b. M'ch 23, 1824; d. Oct. 1, 1854; m. Sam Jonson, but no chil.
4.—**Augustus M.**, b. April 17, 1827; d. Sept. 30, 1855.
5.—**Samuel A.**, b. May 31, 1829; hotel keeper and mercantile man; m. but has no chil.; res. 1235 South 8th street, Phila., Pa.

51 Benjamin Twining, son of **24** John; b. Nov 9, 1797, N. J.; d. M'ch 9, 1883, Crawford Co., Wis., of paralysis; m. Sept. 1, 1823, Mariamna Atkins in Broome Co. N. Y.; b. Aug. 3, 1805, N. Y.; d. Oct. 8, 1871, Wis. Benjamin and family moved to Crawford Co., Wis., M'ch 5, 1855. David, his son, came the year before, returned in the fall, and brought his parents the next spring.

ISSUE:

1.—**Elisha L.**, b. July 23, 1824; d. July 23, 1837.
2.—**Mary E.**, b. April, 1826; d. 1827.
3.—**Nancy**, b. May 1, 1828; d. M'ch 6, 1835.
4.—**Byron**, b. April 4, 1830; d. M'ch 7, 1835.
5.—**108 David M.**, b. July 25, 1832; m. Phebe A. Evans.
6.—**Judith S.**, b. June 6, 1834; d. M'ch 23, 1835.
7.—**Anna S.**, b. Jan. 11, 1836; d. April 30, 1850.
8.—**Rachel R.**, b. June 16, 1838; d. May 11, 1850.
9.—**Ben. H.**, b. July 25, 1840; d. April 22, 1850.
10.—**Adna A.**, b. Sept. 3, 1842; d. May 14, 1851.
11.—**Mariamna**, b. Jan. 29, 1845; m. Charles R. Rounds, Feb. 24, 1861; res. Crawford Co., Wis. *Issue:*
   1. Mary A., b. April 29, 1862.

2. Arthur H., b. Nov. 29, 1865.
3. Rose B., b. April 8, 1867.

12.—**Elisha V.**, b. April 6, 1847; d. May 2, 1850.

52 Mahlon Twining, son of **24** John; b M'ch 20, 1802, N. J.; d. Dec. 26, 1849, in Susquehanna Co., Pa.; froze to death in a snow drift within sight of his own house; farmer; m. Lucy L. Goodseede, b. in Berkshire Co., Mass., Sept. 23, 1801; res. with her son, Henry, in Brookville, Kan.; Bapt.

ISSUE:

1.—**109 Joseph**, b. Jan. 11, 1826; m. Emeline Birdsell.
2.—**Thankful**, b. Oct. 2, 1827; d. M'ch 29, 1858; m. Sol. Tripp; five chil.
3.—**110 Chester P.**, b. M'ch 13, 1829; m. Ann Defan.
4.—**111 Fredrick F.**, b. Oct. 27, 1831; m. Helen F. Payne.
5.—**Franklin**, b. April 10, 1833; d. June 10, 1834.
6.—**112 William F.**, b. Sept. 27, 1834; m. Eleanor Keyes.
7.—**Emma R.**, b. April 10, 1836; d. Sept. 5, 1884; m. Edgar Harper; farmer; res. Binghamton, N. Y. *Issue:*
    1. Fred. B., b. Dec. 8, 1856; m. Belle Robinson; banking business; res. Detroit, Mich.
    2. Carrie L., b. April 5, 1864; teacher; res. Binghamton.
8.—**113 George R.**, b. M'ch 8, 1838; m. Eliz. Carman.
9.—**Jeremiah**, b. April 3, 1840; d. Nov. 19, 1855; farmer; Bapt.
10.—**114 Mahlon J.**, b. Oct. 8, 1841; m. Fanny Gaylor.
11.—**115 Henry L.**, b. July 17, 1843; m. M. C. McCullick.

53 Benjamin Twining, son of **25** Daniel, b. Aug. 30, 1810, Warren Co., N. J.; d. Oct. 29, 1868, Luzerne Co., Penn.; m. Eliz. Lance, b. M'ch 27, 1815; d. May 24, 1886.

ISSUE:

1.—**116 Jessie**, b. April 13, 1834; m. Mary Goodwin.
2.—**117 Eli**, b. Feb. 10, 1836; m. Hannah Taylor.

3.—**Elnora**, b. Nov. 30, 1837; m. M. A. Whitman, b. Nov. 19, 1831. One chil., Eliz., b. Jan. 18, 1855; res. Scranton, Pa.
4.—**Mary J.**, b. Jan. 10, 1843: m. Jno. Jones, b. 1834: railroad employ: res. Scranton.
5.—**118 John**, b. April 27, 1845: m. Nellie Switzer.
6.—**119 William**, b. M'ch 26, 1847: m. Annie D. Gifford.
7.—**Sarah C.**, b. Feb. 25, 1849: m. John Lisk, b. 1853, and has one child, Effie, b. 1876: teamster: res. Scranton, Pa.
8.—**Hannah**, b. Sept. 11, 1850: m. Azor Philo, b. Dec. 10, 1847: railroad employe: res. Scranton. *Issue:*
    1. Lena W., b. Oct. 13, 1871.
    2. Cora B., b. Feb. 3, 1874.
    3. Harry, b. M'ch 4, 1882.
9.—**120 Horace G.**, b. July 25, 1854: m. Minnie Sisco.
10.—**121 Ralph**, b. Sept. 19, 1858: m. Annie Harris.

54 Jacob Twining, son of **25** Daniel, b. about 1816, Warren Co., N. J.; d. 1885, Belvidere, N. J.; m. *first*, Sidney Gano, 1835; d. 1853; m. *secondly*, Mrs. Eliza Townsend, May, 1857; carpenter; wid. res. at Scranton, Pa.

ISSUE:

1.—**Jane**, b. M'ch. 1839: m. Wm. Thatcher, Jan. 1, 1857, by whom she had six chil. He d. 1871: m. *secondly*, John Hunt; res. New Village, Warren Co., N. J.: farmer.
2.—**John**, b. 1841, d. 1862, in the great rebellion: unm.
3.—**122 Samuel**, b. 1843; m. Margaret Rush.
4.—**James**, b. 1845, d. 1864: butcher: unm.
5.—**Sarah**, b. Feb. 4, 1848: m. Jno. W. Knapp, Nov. 9, 1872: carpenter: res. Scranton, Pa. *Issue:*
    1. Louise, b. M'ch 3, 1876.
    2. Herbert L., b. Dec. 28, 1877.
6.—**Mary**, b. 1851, d. 1855.

55 Jacob Twining, son of **26** Joseph, b. Oct. 7, 1770, Wrightstown; d. May 23, 1848, W.; m. Phebe Tucker, May 15, 1793; b. April 26, 1775; d. April 18, 1855, W.; Quakers; farmer.

ISSUE:

1.—**123 Malichi**, b. Aug. 3, 1794: m. Ann Twining.
2.—**Mary**, b. Nov. 2, 1795: d. young.
3.—**Phebe**, b. May 27, 1798: d. about 1831, Wrightstown: m. *Samuel T. Van Horn, who m. again in Bucks Co. and then moved with family to Ohio, 1833. Had two chil. by second marriage. *Issue:*

   1. Jacob, b. April 24, 1818; d. at age of 12 yrs.
   2. Charles, b. Sept. 25, 1819; res. Hancock Co., Ohio.
   3. Isaac, b. June 24, 1821; res. Henry Co, Ohio.
   4. Sarah A., b. Jan. 11, 1823; lives in Mo.
   5. Smith, b. Aug. 26, 1826; res. Mt. Blanchard, Ohio. Aided in getting the Twining family records of Hancock Co, O.

4.—**124 Joseph**, b. Aug. 10, 1800: m. Mary Loosely.
5.—**125 John**, b. M'ch 22, 1802: m. Mary Lambert.
6.—**Sarah**, b. Sept. 9, 1804: d. Feb. 16, 1879: m. Charles Van Horn, 1826. Went to Hancock Co., Ohio, about 1830. *Issue:*

   1. George, b. Dec. 3, 1827; m.; res. H. Co., Ohio.
   2. Moore, b. Jan. 11, 1829; m.; res. H. Co., Ohio.
   3. Robert, b. Sept. 9, 1830; m ; res. H. Co., Ohio.
   4. Mary, b. Feb. 21, 183-; m.; res. Mich.
   5. Martha, b. May 5, 1834; m.; res. Ill.
   6. Harrison, b. Oct. 9, 1836; m.; res. H. Co., Ohio.
   7. Phebe, b. Aug. 7, 1838; m.; res. H. Co., Ohio.
   8. Sarah, b. Dec. 26, 1840; m.; res. H. Co., Ohio.
   9. Charles E., b. April 7, 1843; m.; res. H. Co., Ohio.
   10. John E., b. Jan. 19, 1846; m.; res. H. Co., Ohio.

The above are all prominent and well-to-do people.

*The Van Horn family in America began with Abraham Van Horn, from Holland, who m. Martha Daugan, of Bucks Co., Pa. The only child known was Isaac, b. Nov. 5, 1745; m. Mary Betts, of Bucks Co. (See **11** Samuel Twining.) *Issue:*

   1. Isaac, b. Jan. 25, 1787; d. young.
   2. Sarah, b. Nov. 5, 1789; m. James Moore and raised a large family.
   3. Abraham, b. Jan. 10, 1791; m. Susan Buckman; seven chil.
   4. Samuel T., b. Oct. 1, 1792. (See **55** Jacob Twining.)
   5. John, b. June 25, 1794; res. Pa.
   6. Aaron, b. May 7, 1796; res. Pa.
   7. Martha, b. M'ch 25, 1799; res. Pa.
   8. Charles, b. April 18, 1801. (See **55** Jacob Twining.)
   9. Anna, b. Sept. 19, 1803; unm.

7.—**Hannah**, b. M'ch 12, 1807, d. in Bucks Co.: m. —— Lambert, who d. M'ch 4, 1837, Bucks Co. It has not been revealed that she had chil. or m. second time.

8.—**126 James**, b. Oct. 10, 1808: m. Elizabeth Staley.

9.—**Martha**, b. M'ch 10, 1810, d. in Michigan, 1881: m. Joseph Tucker: chil.: a dau., Elizabeth Gordon: res. Burnip's Corners, Allegan Co., Mich.; second cousin. Moved to Hancock Co., O., and subsequently to Mich. *Issue:*

    1. Sarah, b. Jan. 30, 1830; d. about 1848.
    2. John, b. Sept. 23, 1836; d. in Cal., 1882.
    3. Lydia, b. Aug. 17, 1838; res. Mich.
    4. Elizabeth, b. May 5, 1839; res. Mich.
    5. Jessie C., b. M'ch 5, 1842; d. at age of 25 yrs.
    6. Anna M., b. Aug. 24, 1849; res. Mich.
    7. Margaret, b. Sept. 16, ——; res. Mich.

10.—**127 Jacob**, b. April 12, 1812: m. Elizabeth Adams.

11.—**Lydia**, b. Aug. 4, 1814, d. May 25, 1856: m. Sept. 1, 1836, Wm. Hellyer, b. Dec. 4, 1811, d. M'ch 22, 1885, Penn's Park, Pa. *Issue:*

    1. Harrison C., b. Sept. 25, 1841; d. 1864; m. Elmira Watson; chil.
    2. Hannah E., b. July 9, 1844; m. D. Krusen Harvey; chil.
    3. Howard A., b. Oct. 22, 1845; m. June 28, 1868, Fannie E. Olmsted; physician and prominent citizen; res. Penn's Park; 8 children.

12.—**128 Ralph L.**, b. July 23, 1820: m. Annie Heaston.

56 John Twining, son of 26 Joseph, b. Oct. 21, 1773; d. May 27, 1827, Bucks Co.; m. *first*, Ann, dau. of 14 Eleazer Twining, d. Dec. 5, 1815, aged about 49; m. *secondly*, Elizabeth ——, who d. M'ch 3, 1837, Warwick Tp.

ISSUE BY FIRST WIFE:

1.—A daughter; d. aged 15 years.
2.—**129 Silas**, b. April 26, 1802, Bucks Co.; m. Letitia Harrold.

57 Joseph Twining, son of 26 Joseph; b. Nov. 8, 1780; d. April 11, 1860, Newtown, Pa.; m. *first*,

Mary Tucker, his first cousin and sister to his brother,
55 Jacob's wife, b. July 27, 1777; d Jan. 12, 1844;
m. *secondly*, Elizabeth Borroughs.

ISSUE:

1.—**George W.**, b. 1806; d. April 2, 1874, Findlay, Ohio; m. Evaline Scarborough and moved to Hancock Co., O., where wid. res.; Congregationalists; no issue.
2.—**Susanna**, b. 1808; d. at Christiana Hundred, Delaware, Aug. 16, 1886: m. Oliver P. Ely, of Wilmington, Del. *Issue:*
    1. George (twin); farmer.
    2. Thompson (twin).
    3. Louisa, m. —— Lyman.
    4. Emma, m. —— Talley; res. Wilmington, Del.
3.—**130 Jonathan R.**, b. Nov. 19, 1809; m. Susan Balliel.
4.—**Antoinette L.**, b. 1813, d. Dec. 22, 1885, Lambertsville, N. J.; m. Ira T. Johnson, of Newtown Tp. One son, James T., living in L., N. J., and one, Wm. H., res. N. Y. City.
5.—**Mary**, b. 1815, d. Dec. 5, 1855; m. Mahlon Reeder (son of Abraham), b. April 10, 1806. A strong, healthy man; res. Penn's Park, Pa. *Issue:*
    1. Geo. W., b. Dec. 18, 1833; watchman; res. Philadelphia, Pa.
    2. Lewis A., b. Sept. 28, 1835; builder; d. Aug. 26, 1884, Helena, Mont.
    3. Huston T., b. Sept. 28, 1836; carpenter and engineer; res. Helena.
    4. Maria Louisa, b. April 2, 1838; d. June 1, 1880, Danville, Pa.
    5. Abraham R., b. Sept. 18, 1842; d. April 22, 1883, Phila.; engineer.
    6. Joseph, b. Jan. 19, 1848; d. April 7, 1853, Penn's Park.
    7. Willes W., b. Sept. 7, 1850; real estate broker; res. Philadelphia.
    8. Stephen, b. Nov. 10, 1855; d. Jan. 9, 1856, Penn's Park.
6.—**Mercy M.**, b. 1818, d. M'ch 28, 1854; m. Charles Hart, Sept. 19th, 1839; a carpenter by trade; res. Penn's Park, Bucks Co., Pa. *Issue:*
    1. Harrison, b. Aug. 9, 1840; invalid; res. Newtown, Pa.
    2. Joseph T., b. April 4, 1842; carpenter; res. Solebury, Pa.
    3. Samuel T., b. Dec. 16, 1843; d. Feb. 11, 1844.
    4. Mary Jane, b. April 29, 1845; m. Nelson Heston; farmer; res. Wrightstown, Pa.

5. Elizabeth E., b. Jan. 14, 1847; m. Howard Barrvis; tailor; res. Morrisville, Pa.
6. Albert H., b. May 1, 1849; d. Sept. 8, 1874.
7. Amanda P., b. June 2, 1851; m. Charles Swope, laborer; res. Lambertville, N. J.
8. Susanna E., b. Nov. 5, 1852; d. Aug. 18, 1854.
9. Antoinetta L., b. Jan. 31, 1854; d. Aug. 14, 1854.

58 Watson Twining, son of **27** Silas; b. Nov. 20, 1797, Bucks County; d. April 13, 1847; buried in Warminster burial grounds; m. Margaret, dau. of Joseph and Rebecca Hallowell, of Moreland, Mont. Co., Dec. 6, 1821; d. 1888; b. M'ch 27, 1802; Friends.

ISSUE:

1.—**Mary**, b. Nov. 4, 1822; d. Sept. 22, 1825, Wrightstown.
2.—**131 Hallowell S.**, b. April 5, 1824; m. Jane Williams.
3.—**Rebecca H.**, b. Dec. 11, 1825; living in Philadelphia; single.
4.—**Amos W.**, b. Dec. 8, 1828; res. Phila.
5.—**Elizabeth**, b. May 6, 1830; m. Wm. J. Kirk, May 22, 1856; res. Warminster Tp., Bucks Co.; no issue.
6.—**132 Elias B.**, b. Sept. 26, 1832; m. Charlotte Tyson.
7.—**Mary**, b. M'ch 23, 1835; d. April 17, 1856, Warminster.
8.—**Joshua D.**, b. Sept. 10, 1838; m. Eliz. Patterson, Oct. 6, 1874; one child, Edith J., b. Jan. 22, 1878; res. Philadelphia.
9.—**Alice W.**, b. Sept. 10, 1840; d. Aug. 10, 1844, W.
10.—**Anna**, b. Oct. 9, 1843; m. George Shoemaker, Aug. 10, 1874; d. July 15, 1879; he d. M'ch 8, 1882, Warminster Tp.
11.—**Alice W.**, b. Jan. 6, 1846; d. May 25, 1846, W.

59 Silas Twining, son of **27** Silas; b. M'ch 27, 1807, Warminster, Bucks Co.; d. Aug. 19, 1847, Wrightstown; Friend; farmer; m. *first*, Martha Simpson, d. Aug. 23, 1840; m. *secondly*, Amanda K. Simpson (no relation), dau. of James and Mary, Dec. 7, 1843: Presbyterian.

ISSUE by First Wife:

1.—**Ruthanna**, b. Sept. 25, 1837.
2.—**Elmira W.**, b. Jan. 25, 1839; educated Claverack Sem., N. Y.; Friend; res. Newtown, Pa.; single.

ISSUE by Second Wife:

3.—**Samuel W.**, b. Dec. 14, 1844; m. Marion, dau. of Jonathan B. and Eliza R. Wright, Nov. 12, 1868; one child, Anna St. John, b. Aug. 2, 1880; printer; res. Brooklyn, N. Y.
4.—**Mary**, b. July 8, 1846; d. young.
5.—133 **Silas**, b. M'ch 21, 1848; m. Annie Vanartsdalen.

60 William Twining, son of **28** David, b. April 13, 1797, Bucks; d. Dec. 22, 1856, B.; m. Rebecca Riley, of N. J., 1826: b. Feb. 2, 1800: res Trevose, Bucks Co.: family Quakers.

ISSUE:

1.—**Eleazer**, b. Feb. 22, 1828; m. Hannah Lacy, May 14, 1872, blacksmith; M. E. Ch.; res. Frankford, Pa.; one child, Naomi, b. July 14, 1873.
2.—**Reuben L.**, b. Oct. 11, 1829; d. Nov. 28, 1865; single.
3.—134 **Uriah R.**, b. Oct. 16, 1831; m. Juliann Vanartsdalen.
4.—**Phebe A.**, b. M'ch 20, 1834; m. Septimus Tucker, Aug. 29, 1860; res. Bucks Co.
5.—**Amy L.**, b. Aug. 25, 1836; m. Thos. Amos, Nov. 20, 1856; res. near Annapolis, Md.
6.—**Rebecca S.**, b. July 24, 1841; d. May 1, 1844.
7.—135 **William W.**, b. Jan. 17, 1844; m. Mary A. Van Horn.

61 Isaac Twining, son of **28** David, b. Aug. 8, 1802, Bucks; d. Nov. 8, 1882, Harford Co., Md., where he moved to from B., 1845; farmer; Quaker; children mostly of same persuasion; m. Nov. 15, 1827, Ann L., dau. of Dan. and Mary Hallowell, M'ch 19, 1803: d. Aug. 21, 1877.

ISSUE:

1.—**136 D. Hallowell,** b. Aug. 29, 1828; m. Alice P. Baynes.
2.—**Martha E.,** b. Aug. 20, 1830; single; farmer; res. Upper Cross Roads, Harford Co., Md.
3.—**137 Horace B.,** b. Sept. 15, 1832; m. Fannie Ashton.
4.—**Isaac T.,** b. Dec. 7, 1834; 1857 in Kan. and Mo.; 1859-60 Miss., from whence he drifted into the *Rebel army*, where he remained during the war. In 1885 was living in Senior, Tex., where he m. a wid. with two chil. She d. and his whereabouts is not known.
5.—**138 B. Franklin,** b. Oct. 12, 1837; m. Mary C. Nippes.
6.—**Caroline,** b. M'ch 7, 1840; m. Wm. D. Bartleson, Feb. 17, 1865, from Del. Co., Pa. Settled in Pleasantville, Harford Co., Md., 1853. *Issue:*

    1. Anna M., b. April 12, 1867.
    2. Martha L., b. April 19, 1869.
    3. William D., b. Dec. 23, 1879.

7.—**Robert Barclay,** b. Feb. 16, 1843; killed at the second battle of Bull Run, Aug. 29, 1862; disowned by Friends.

62 Thomas Twining, son of 28 David;

b. Feb. 16, 1808, Bucks: d. Dec. 6, 1872, Bloomington, Ill.: m. *first*, Sarah A. Bean, M'ch 8, 1832: b. Oct. 18, 1811, d. July 17, 1845: dau. of Isaac and Hannah: m *secondly*, Mrs. Alcinda E. Randolph, dau. of Ben. Cundiff, Jan. 28, 1849, b. Oct. 15, 1816, Vir.: res. B.: M. E. Ch.

From the McLean Co., Ill., Hist. we extract the following:

Thomas Twining, farmer and stock raiser; early pioneer of McLean Co.; attended coll. 5 years, two of which were devoted to medicine. In 1836 emigrated to Ill. from Bucks Co., Pa.; entered 300 acres in Old Town Tp.; took active interest in politics, being an old line Whig and then Rep. Held township, school and justice offices. At one time saved the county $10,000. Borrowed money to enter his first claim, but at his death owned 500 acres; Quaker.

ISSUE BY FIRST WIFE:

1.—**Mary E.,** b. June 2, 1833; m. Archibald Campbell, Feb.

6, 1851; retired farmer, Bloomington, Ill.; M. E. Ch. *Issue:*

1. Mary B., b. Feb. 21, 1852; m. James Weidner; chil.
2. Elizabeth, b. Dec. 19, 1856; m. P. W. Gregory.
3. Thomas A., b. May 2, 1859; m. M. E. Noggle.
4. Franklin E., b. Dec. 9, 1861; d. Oct. 13, 1886.
5. Howard A., b. Aug. 2, 1865; d. Jan. 31, 1882.
6. Chas. E., b. Oct. 23, 1867.
7. Nellie A., b. May 24, 1871.

2.—**Louisa E.**, b. M'ch 3, 1835; m. Peter C. Jacoby, Feb. 15, 1866; farmer; M. E.; res. Holden, Ill. *Issue:*

1. Thomas H., b. Nov. 12, 1866.
2. Daniel A., b. Nov. 1, 1868.
3. Dellie M., b. Aug. 12, 1871.
4. Franklin E., b. June 7, 1874.

3.—**Martha**, m. John Kendall; farmer; res. Farmer City, Ill.
4.—**Sarah A.**, b. April 22, 1845; d. July 25, 1845.
5.—**Thomas C.**, b. May 24, 1850; d. Oct. 8, 1851.
6.—**Isaac**, b. Jan. 4, 1856; d. April 21, 1856.
7.—**139 Charles H.**, b. Nov. 12, 1853; m. Mary A. Savidge.

63 Jacob Twining, son of **29** John, b. Aug. 6, 1806, Wrightstown; d. Sept. 8, 1882, Boscobel, Wis., where he moved to from Phila. about 1863; m. Rachel Ryan, July 15, 1835, b. Aug. 11, 1814; a native of Phila., Pa.; stone mason and miller; Quaker in belief; wid.: res. Boscobel: M. E. Ch. Children born in Wrightstown.

ISSUE:

1.—**Allen**, b. May 18, 1836; m. Joshua W. Watson, June 20, 1853; res. Boscobel; b. July 29, 1826; painter and d. Aug. 15, 1881. *Issue:*

1. Sarah C., b. Aug. 11, 1856; m. Jno. James; res. Chillicothe, Mo.
2. Hannah R., b. April 18, 1859; m. Joseph Glynn; res. Campbell, Dak.
3. Jacob T., b. Aug. 23, 1861; m. Eliz. Gribble; res. Boscobel.
4. Isaac H., b. Feb. 18, 1865; res. Campbell, Dak.
5. Leuella G., b. Sept. 23, 1867; m. Levi Sanborn; res. B.
6. Mabel C., b. Sept. 19, 1872.
7. Laura M., b. May 2, 1875.

2.—**140 John**, b. Oct. 10, 1839; m. Kath. S. Frankinfield.
3.—**141 Isaac H.**, b. Feb. 23, 1845; m. Mary A. Whit.
4.—**Susanna**, b. Aug. 7, 1852; m. Monroe Pidcock, Dec. 22, 1869, b. Jan 25, 1850, Wrightstown; farmer. *Issue:*
   1. George, b. May 23, 1876.
   2. Charles, b. Nov. 29, 1881.
   3. Mertie, b. Aug. 17, 1886.

64 Abbott C. Twining, son of **29** John, b. Nov. 29, 1810; d. June 6, 1882; m. June 20, 1832, Maria Warner, b. Jan 12, 1810; d. M'ch 9, 1888; farmer; res. Bucks Co., Penn.

ISSUE:

1.—**Martha E.**, b. Sept. 3, 1833; m. Jos. W. Worthington, Oct. 12, 1857; farmer; res. Wrightstown, Pa. *Issue:*
   1. Wm. A., b. June 26, 1859; m. Sarah A. Slack.
   2. Geo. M., b. Feb. 17, 1861.
   3. Anna M., b. Sept. 10, 1863.
   4. Asenith, b. July 13, 1868.
   5. Martha, b. June 6, 1879.
2.—**142 John Warner**, b. June 11, 1837; m. Mary E. Briggs.
3.—**143 Thomas Chalkley**, b. Feb. 12, 1844; m. Mary E. Kirk.
4.—**Rachel A.** (twin), b. May 17, 1848; m. Achilles Blaker, M'ch 16, 1871 (deceased); one child, Matilda, b. Dec. 10, 1871; teacher in Friends' school. She owns a farm at Wrightstown, Pa.
5.—**Sarah E.** (twin), b. May 17, 1848; m. John Kirk, Oct. 8, 1868; farmer, near the forks of Neshamney creek; P. O. Rush Valley, Pa. *Issue:*
   1. Emily A., b. April 17, 1869.
   2. Abbott C., b. Aug. 23, 1873.
   3. Wm. R., b. Sept. 2, 1875.
   4. Mary A., b. Sept. 19, 1880.

65 Isaac H. Twining, son **29** John, b. Oct. 21, 1812, d. March 22, 1856, Philadelphia; m. Phebe, dau. of Thomas and Eliz. Megadigan, May 7, 1846,

d. M'ch 27. 1854, Bucks Co. Isaac was a chairmaker by trade, living in Bucks Co., near Wrightstown.

ISSUE: One child.

—**144** David R., b. Nov. 29, 1851; m. Hannah Kyle.

66 Jessie B. Twining, son of **30** Jacob, b. Sept. 25, 1817; m. Hannah Beans, Dec. 14, 1848; b. June 9, 1820; farmer; res. Richborough, Bucks Co. Pa.

ISSUE:

1.—**Sarah B.**, b. Sept. 14, 1849; m. Wm. Smith, Jan. 7, 1874. Children: Hannah and Mary Alice; res. Richborough.
2.—**William**, b. April 2, 1851; d. inf.
3.—**Ruth**, b. Nov. 21, 1852.
4.—**Jacob**, b. May 11, 1855; res. Newtown, Pa.
5.—**Rachel**, b. April 1, 1860; d. inf.
6.—**Albert H.**, b. Oct. 1, 1861; m. Margaret Hoagland, Nov., 1885. In banking business, Asbury Park, N. J.; child, a dau., Jessie W., b. Nov. 28, 1888.

67 Henry M. Twining, son of **30** Jacob, b. Jan. 4, 1820; m. Oct. 11, 1849, Elizabeth Longshore, who d. July 19, 1884, aged 57 years, 1 day; Quakers; res. Philadelphia.

ISSUE:

1.—**145** Howard L., b. July 13, 1850; m. Mary Cooper.
2.—**Mary S.**, b. Jan. 31, 1854; d. June 21, 1855; drowned.
3.—**Thomas M.**, b. M'ch 29, 1857; m. Ellen E. Woods, June 8, '81; pork business; res. Philadelphia.
4.—**Allen H.**, b. Aug. 1, 1859; m. Achsah Paul, Oct. 18, 1882; machinist and store-keeper for Reading R. R. Co.; res. Philadelphia.

68 Cyrus B. Twining, son of **30** Jacob, b. Sept. 25, 1827; m. Sarah A. Atkinson, Oct. 7, 1851; b. Jan. 19, 1825. Disowned from O. Friend for m. Hixite Friend; farmer; res. Pineville, Pa. Also in pork business, Phila.

ISSUE:

1.—**146 Jonathan A.**, b. Sept. 10, 1852; m. Belle Warner.
2.—**Thomas O.**, d. inf.
3.—**Ellen S.**, b. M'ch 18, 1854; m. Stephen K. Cooper, Nov. 25, 1875; pork business; res. Philadelphia. One child, John W., b. May 18, 1880.
4.—**William D.**, d. Dec. 30, 1857.
5.—**147 Wilmer A.**, b. April 17, 1865; m. Lottie Vandegrift.

69 Amos H. Twining, son of **31** David, b. May 31, 1820: m. Mary Tomlinson, 1843: farmer: res. Richborough, Bucks Co., Pa.

ISSUE:

1.—**Geo. W.**, b. Nov. 16, 1843; farms his father's farm; single.
2.—**148 William H.**, b. Feb. 8, 1845; m. Mary C. Eckert.
3.—**David**, b. Feb. 1, 1847; d. Sept. 13, 1866.
4.—**149 John**, b. Dec. 11, 1849; m. Mary E. Slack.
5.—**Paul Milton**, b. Jan. 21, 1851; single; res. Richborough.
6.—**Sallie E.**, b. M'ch 23, 1853; teacher; educated Millersville Normal School; single; res. Richborough.
7.—**Wamsley R.**, b. M'ch 30, 1855; master builder and contractor; single; res. Richborough, Pa.
8.—**Mary R.**, b. M'ch 22, 1856; single; res. R.

70 George Twining, son of **31** David, b Oct. 24, 1823, d. Jan. 28, 1872, Lancaster, Pa.; m. Anna C. Eberman, of L., Feb. 10, 1848: b. Sept. 25, 1826: whip maker: M. E. Church.

ISSUE:

1.—**Edmund C.**, b. Jan. 19, 1849; m. Mary E. Eleman, of Philadelphia, May 5, 1875; printer; Camden, N. J.
2.—**John E.**, b. July 10, 1851; d. April 31, 1852.
3.—**Maria H.**, b. April 20, 1854; res. Lancaster.
4.—**David**, b. Nov. 25, 1856; d. inf.
5.—**James P.**, b. Nov. 13, 1858; res. Lancaster, Pa.; clerk.
6.—**Elizabeth A.** b. Nov. 25, 1861; m. John A. Bechtold, April 25, 1883; d. M'ch 24, 1861; res. Lancaster, Pa. *Issue:*
   1. Paul E., b. Jan. 24, 1884.
   2. Edna R., b. Nov. 2, 1886.

7.—**William**, b. July 5, 1864; d. Nov. 6, 1864.
8.—**Rachel**, b. June 11, 1866; m. Wilmer E. Barton, Feb. 22, 1883; b. Jan. 29, 1864; res. Lancaster. *Issue:*
    1. Robert E., b. May 7, 1884.
    2. Meriam, b. Nov. 13, 1886.
    3. David T., b. M'ch 1, 1888.

71 Croasdale Twining, son of **32** David, b. May 7, 1803; d. of paralysis of the heart, Feb. 16, 1888, Wrightstown, Pa.; buried at Ivy Hill Cemetery, Phila.; m. Mary, dau. of Isaac and Sarah Kirk, Oct. 27, 1833: lived for many years at Soldiers' Grove, Montgomery Co., Pa.: farmer: wid. res. Tacoma, Wash.

ISSUE:

1.—**Louisa**, b. M'ch 18, 1835; res. Tacoma, Wash.
2.—**150 Edwin**, b. Dec. 1837; m. Hannah A. Iredell.
3.—**G. Chapman**, b. 1839; d. Dec. 29, 1882.
4.—**Alfred**, b. 1841; d. 1844.
5.—**Elizabeth**, b. 1843; d. 1846.
6.—**Margery**, b. May, 1846; m. Wm. Kite, Jr., of Germantown, Pa.; res. Philadelphia, Pa.
7.—**Caroline**, b. Aug. 1855; m. William Sharps, who served in the Rebellion; one child, Magdalena; res. Tacoma, Wash.

72 Stephen Twining, son of **32** Jacob, b. June 25, 1805, d. Aug. 1, 1882, Langhorne, Bucks Co., Pa: m. Sarah A., dau. of Jno. Warner, 1832, b. Oct. 12, 1812: carpenter by trade.

ISSUE:

1.—**John W.**, b. July 28, 1833; moved from Bucks Co. to Ill., and then to Iowa, where he m. Maggie Oglebee; res. Griswold, Cass Co., Iowa. *Issue:*
    1. Sallie, b. Aug. 20, 1873.
    2. Hannah E., b. Feb. 14, 1875.
    3. Fannie A., b. Dec. 18, 1877.
    4. Mary D., b. Nov. 9, 1879.
    5. Eva T., b. Sept. 18, 1881; d. June 18, 1883.
    6. Minnie, b. Aug. 13, 1883.

THE TWINING FAMILY.

1.—**Mary A.**, b. Oct. 8, 1835; m. Jonathan Hibbs; she d. M'ch 12, 1872. *Issue:*
   1. Jonathan, b. June 26, 1856.
   2. Eliza T., b. Jan. 1, 1859.
   3. William, b. Aug. 11, 1861.
   4. Cordelia, b. Oct. 2, 1866.
3.—**William**, b. Feb. 17, 1838; d. Sept. 15, 1843.
4.—**Martha B.**, b. July 27, 1840; m. Geo. McDonald. *Issue:*
   1. Eva, b. Nov. 13, 1862.
   2. Emma, b. July 23, 1866.
   3. Wm. W., b July 24, 1869.
   4. Geo. F., b. June 10, 1872.
   5. Sarah E., b. Feb. 26, 1875.
   6. Warner T., b. May 31, 1876.
5.—**Charles C.**, b. Feb. 6, 1843; d. Oct. 28, 1850.
6.—**Croasdale**, b. May 27, 1846; d. Nov. 3, 1864, in war of Rebellion.
7.—**Sarah B.**, b. M'ch 31, 1853; m. Elwood Stephens. *Issue:*
   1. Eugene F., b. April 8, 1875; d. Aug. 8, 1875.
   2. Ben. F., b. Feb. 25, 1877.
   3. Ida E., b. April 8, 1879; d. April 20, 1880.
   4. Hoagland B., b. Dec. 14, 1880; d. June 16, 1883.
   5. Emma M., b. Feb. 21, 1882.
   6. Elwood, b. Dec. 12, 1884; d. May 25, 1885.

73 Alexander Catlin Twining, son of **33** Stephen, b. New Haven, Ct., 1801; d. Nov. 22, 1884, N. Haven; m. Harriett Kinsley, of West Point, N. Y., M'ch 2, 1829; d. 1871, Grad. Yale 1820. Civil engineer; Prof. Math. Middlebury Coll., Vt., some years; classmate and intimate friend of President Woolsey and Rev. Leonard Bacon, D.D. Associate with Profs. Silliman and Olmsted in scientific observations.

From the New York *Independent* is the following extract:

"The death of Prof. A. C. Twining ends a long life of varied and brilliant achievements, and which was even richer and more beautiful in richness and fruitfulness of Christian character. Prof. Twining is known among astronomers as the author of the cosmic theory of the meteors. As a civil engineer he was engaged as chief or controlling engineer on every line running out of New Haven on the northern roads through Vermont, on the Lake Shore, the Cleveland, Columbus and Pittsburg and various roads out of Chicago, including the Rock Island and the old Milwaukee line.

"As an inventor he pioneered to a successful result the industrial manufacture of artificial ice.

"For nine years he served as Professor of Math. and Astronomy in Middlebury Coll., and while thus residing in Vt. was active in the Temperance Reform, into which he entered with energy as the Chairman of the State Temperance Committee. In political matters he took deep interest as one of the promoters of the original movements which issued in the foundation of the Republican party.

"He was one of the projectors of the famous Conn. letter to President Buchanan. He was deeply interested in constitutional questions, a study which culminated in his lectures on the Constitution of the United States in Yale Law School. In questions of theology and philosophy he was at home and discussed them with bold vigor and subtile ingenuity. On his friends, the beauty of his face and head, the striking and winning courtesy of his manner, the simplicity of his Christian character made a lasting impression, while few that ever met him even casually have failed to notice that to him it was given to invite and receive the spiritual confidence of others, and to give them solid and permanent assistance where there are few to attempt it and still fewer to succeed."

ISSUE:

1.—**151 Kinsley**, b. July 18, 1832; m. Mary K. Plunkett.
2.—**Harriett A.**, b. Dec. 27, 1833; unm.
3.—**Theodore Woolsey** (twin), b. Sept. 4, 1835; d. Aug. 14, 1864.
4.—**Sutherland Douglas** (twin), b. Sept. 4, 1835; grad. Sheffield Scientific School; Yale, 1859 (?); physician; res. Chicago, Ill.; m. Gertrude Tenny, who d. 1880; no issue surviving.
5.—**Sarah Julia**, b. Nov. 9, 1837, "a bright and resolute woman;" single.
6.—**Mary Elmira**, b. April 23, 1840; m. A. D. Gridley, of Clinton, N. Y.; he d. 1876; no chil.; she resides in N. Haven.
7.—**Eliza Kinsley**, b. June 19, 1843.

74 William Twining, son of **33** Stephen, "b. New Haven, Ct., Dec. 9, 1805. Graduated Yale Coll. 1825. Member of Yale Theological Sem., 1826, and Andover Theological Sem., 1827. Ordained, Great

Falls, N. H., 1830, Jan. 6. Acting pastor there 1830, Jan. to 1831. Installed, Appleton Street (now Eliot) Church, Lowell, Mass., 1831; dismissed Aug. 25, 1835. Principal Female Seminary, Madison, Ind., 1836–43. Professor of mathematics, natural philosophy and astronomy, Wabash College, 1843–54. Without charge, Crawfordsville, Ind., 1854–59. Acting pastor, Beardstown, Ill., 1859–63. Without charge, St. Louis, Mo., until death. Published Antiphonal Psalter and Liturgies, 1877. Married June 1, 1830, Margaret Eliza, daughter of Horace and Catharine (Thorn) Johnson, of New York City, who d. Oct. 15, 1873. * *
* * * * Died of paralysis in Laclide, a suburb of St. Louis, 1884, June 5, aged 78 years, 5 mo. and 26 days."—(Congregational Year Book, 1885.)

"These two brothers (**73** and **74**) were men of strong and cultured minds and of perfectly upright characters; they were always physically vigorous."

ISSUE:

1.—**Almira Catlin**, b. Great Falls, N. H., July 22, 1831; m. 1851, Rev. Charles H. Marshall; Cong. Ch.; she d. 1865 (?) Had six children, all boys, of whom only two survive (1884):

   1. Charles H., b. about 1860; m. 1876, Ida Porter; ins. business; res. Crawfordsville, Ind.
   2. Edward H., b. July 17, 1863; grad. Wabash Coll., 1884; d. Nov. 8, 1887.

2.—**152 Edward Henry**, b. Lowell, Mass., Oct. 3, 1833; m. Harriett Sperry.
3.—**William Alex.**, b. 1835; d. 1836.
4.—**Cathrine Ann**, b. Madison, Ind., M'ch 1, 1837; m. 1863, Charles D. Moody, lawyer, Beardstown, Ill.; res. St.

Autograph of 74 Yours &c. William Twining

Louis, Mo. She was teacher in Cleveland, O., Fem. Sem., 1856-9; Congregational Ch. *Issue:*

    1. Harriett, b. 1865; d. 1866, St. Louis.
    2. Catharine, b. 1867.
    3. Ethelwynne, b. 1870(?); d. 1884, Webster Grove, Mo.
    4. Constance, b. 1875, St. Louis.
    5. Mark, b. 1877(?); St. Louis.

5.—**William Johnson**, b. Madison, Ind., Aug. 2, 1839. Grad. U. S. Military Academy, 1863, and promoted to the rank of First Lieutenant in the Corps of Engineers same date. Served throughout the Civil War in the departments of of the Cumberland and Ohio, as Assistant and Chief Engineer, participating in the battles of Franklin and Nashville and the invasion through Georgia. "For gallant and meritorious services in action during the Rebellion," received rank Major United States Army, M'ch 13, 1865. Astronomer Northern Boundary Survey, 1872-76, and other military duties. In 1878 appointed Engineer Com. of the Dis. of Columbia, having received the rank of Major of Engineers, 1877, in which capacity he distinguished himself with honors. He was considered one of the most accomplished and capable engineer officers in the army. A man of fine personal qualities and sterling integrity. Died of pleurisy, May, 1882; buried with military honors at West Point. Was never married.

6.—**Helen Eliz.**, b. 1841; a beautiful woman of fine musical talent; res. St. Louis, Mo.

7.—**153 Charles Osmond**, b. Sept. 28, 1845; m. Anna Campbell.

8.—**Mary Evelyn**, b. 1849; res. St. Louis, Mo.

75 Alfred C. Twining, son of **34** William, b. Oct. 8, 1804, Tolland; d. Aug. 31, 1883, Lansingburgh, N. Y.; m. *first*, Marietta Hamilton, Dec. 15, 1834; d. Sept. 8, 1841, T.; sister of his brother Stephen's wife; m. *secondly*, Mary F. Barton, b. April 1, 1818, L.; d. May 26, 1886, Troy, N. Y.; merchant; Congregationalist.

THE TWINING FAMILY.

ISSUE BY FIRST WIFE:

1.—**Maria**, m. J. B. Shepard, of Chicago, Ill., and d. 1865.
2.—**Helen**, m. J. B. Shepard, Secretary Department of Police, res. 60, 23d street, (above).
3.—**154 George A.**, b. M'ch 15, 1841, L.; m. Jennie Byers.

ISSUE BY SECOND WIFE:

4.—**William Barton**, b. Jan. 18, 1846: d. Sept. 13, 1848.
5.—**Charles Barton**, b. July 6, 1849: d. May 24, 1867.
6.—**Alfred F.**, b. Dec. 19, 1851: d. M'ch 3, 1854.
7.—**Francis Barton**, b. Sept. 3, 1856. Shirt and collar business: res. Troy, N. Y.: m. Dec. 12, 1889, Nomina Newcomb, dau. of Dr. Daniel D. Bucklin.

76 Alexander Twining, son of **34** William, b. Dec. 25, 1814, T.; d. Feb. 28, 1862, T.; m. Laura J. Tinker, Sept. 22, 1841, b June 14, 1821, T. He was a farmer and lived on the old homestead during life; Cong.; his wid. m. Wm Humphreys, of Middleburgh, Ohio, 1866. She became a wid. again, 1884, and resides with his chil.

ISSUE:

1.—**Emergene L.**, b. Oct. 16, 1843: m. James R. Irwin, Nov. 27, 1872, b. M'ch 9, 1849: farmer: res. Norton Center, Ohio. *Issue:*
    1. Allen T., b. Nov. 2, 1873.
    2. Harry H., b. May 27, 1875; d. Jan. 13, 1877.
    3. James R., b. Feb. 21, 1880.
    4. Nettie C., b. May 24, 1882.
    5. Wm. L., b. July 2, 1884.
2.—**155 William F.**, b. Nov. 18, 1851: m. Eva M. Carpenter
3.—**Cora F.**, b. Feb. 12, 1854: m. Alfred H. Taylor, Nov. 5 1876: farmer: res. Berea, Ohio. *Issue:*
    1. Winnie, b. April 13, 1878.
    2. Pearl A., b. M'ch 2, 1882.
    3. Clyde A., b. Oct. 19, 1884; d. April 16, 1885.
    4. Laura J., b. May 1, 1886; d. Aug. 5, 1886.

77 William Twining, son of **35** William, b. June 14, 1789, T., d. Aug. 8, 1883, So. Rutland, N. Y.; m. April 27, 1813, Ovanda Fowler, d. May 9, 1855. He entered Williamstown Coll. 1808; soon after leaving coll. he made a trip on horseback to Licking Co., Ohio, and Jefferson Co., N. Y., with his uncle, **38** Lewis Twining and bought of him several hundred acres of land in Champion, Jeff. Co., N. Y., to which he removed from T. in the spring of 1818. "He acquired a competence and gave to each of his four sons and three dau. a liberal academic education. Was esteemed for his honesty, liberality and integrity. A devout mem. of Pres. Ch., wherein he was nurtured, until about 1826, when he became a Universalist, his wife being in accord with him Retained his mental powers to the end. When called out for a speech on his 90th birthday, he astonished the assembled guests in a happy and appropriate manner."

ISSUE:

1.—**Susannah**, b. Feb. 14, 1814: m. Rev. J. H. Whelpley, Oct. 15, 1833: m. *secondly*, —— Thompson: res. Mullikin, Michigan. *Issue:*
   1. Jerome Twining, M. D., b. Oct. 18, 1834; res. Cobden, Ill.
   2. Solon R., M. D., b. Feb. 13, 1837; res. Grand Ledge, Mich.
   3. Ferdinand, b. Sept. 19, 1842; res. Hoytsville, Mich.
   4. Cecilia, b. May 17, 1845; res. Portland, Mich.
   5. Byron I., b. Jan. 16, 1851; res. Hoytsville.

2.—**156 John**, b. Jan. 6, 1816: m. Eveline R. Smith.

3.—**Lucinda A.**, b. May 6, 1818: m. *first*, Joab Miller, Feb. 16, 1840, d. 1843: m. *secondly*, Samuel Smith, 1845, divorced 1855: m. *third*, John Mills, Nov. 10, 1861, d. July 8, 1862: m. *fourth*, G. W. Adams, M'ch 10, 1886: res. Brodhead, Wis. *Issue* by first husband:
   1. Overa V., b. Feb. 5, 1841. *Issue* by second marriage:

    2. Corydon Twining, b. Jan. 21, 1847; d. June, 1866, run over by his wagon. Served in the late war.
    3. Milo E., b. Nov. 22, 1851; res. Republican City, Neb.

4.—**157 William F.**, b. Aug. 17, 1820: m. Martha M. Taylor.
5.—**158 Alfred W.**, b. Sept. 3, 1822: m. Jennette Fargo.
6.—**Milo S.**, b. So. Champion, N. Y., Dec. 10, 1826: removed to Wis., 1854: m. Kate A. Rockwood, Dec. 11, 1860; justice peace, dairyman and importer of stock; res. Brodhead, Wis. *Issue:*
    1. Lillian, b. July 30, 1862; m. Fred. A. Mitchell, Jan. 1, 1880; one child, son, b. M'ch 27, 1884; res. Dubuque, Iowa; commercial traveler.
    2. Jessie, b. July 28, 1863; d. Sept. 30, 1864.
7.—**Marietta O.**, b. Oct. 27, 1829; m. *first*, J. W. Smith; m. *secondly*, Silas Weller, Sept. 10, 1860, who d.; res. South Rutland, N. Y. *Issue* by first m.:
    1. Leonora, b. Feb. 27, 1851; m. Oct. 12, 1870, David Waldo. By second husband:
    2. Birdie, b. Dec. 7, 1862; d. M'ch 22, 1864.
    3. Ettie, b. Feb. 17, 1867; m. Frank Stockwell; res. So. R.

78 Elijah Twining, son of **35** William, b. Aug. 25, 1792, T.; d. Nov. 5, 1872; farmer. Owned 400 acres in T.; Presbyterian; m. *first*, Almira More, May 1, 1816; d. July 2, 1870, aged 75 yrs; m. *secondly*, Fidela L. Rogers.

· ISSUE:

1.—**Harriett A.**, b. M'ch 31, 1817; m. Austin H. Ramson, Jan. 1, 1845; res. West Hartford, Ct. *Issue:*
    1. Almira C., b. M'ch 24, 1846; d. Sept. 8, 1863.
    2. Julius E., b. M'ch 19, 1848; d. Oct. 11, 1852.
    3. Ausbert A., b. Aug. 12, 1849; m. Jane Beacher, Jan. 23, 1881.
    4. Lawrence B., b. Oct. 23, 1850; m. Louisa Deny.
    5. John H., b. M'ch 24, 1852.
    6. Julia A., b. May 25, 1856; m. Wm. Fuller, M'ch 20, 1876.
    7. Susan A., b. April 5, 1860; m. Geo. A. Fuller, Nov. 24, 1882. she d. July 24, 1885.
2.—**159 Elphonzo**, b. June 8, 1818; m. Eliza A. Cone.
3.—**160 Joseph**, b. April 23, 1820; m. Henrietta M. Talcott.
4.—**161 Orlandon**, b. Sept. 30, 1821; m. Lucy E. Irvin.

5.—**Eliza A.**, b. Dec. 15, 1822; m. Joseph Kenyon, of Otis Mass., Nov. 25, 1847; farmer and postmaster (1886). *Issue:*
   1. Mary Jane, b. Aug. 6, 1849.
   2. Myra, b. June 16, 1854; m. and has one child.

6.—**162 Samuel M.**, b. Feb. 9, 1824; m. Harriett Gates.

7.—**Bevel**, b. May 8, 1826; m. Melinda E. Brown, Aug. 21, 1871; d. May 27, 1889, aged 66 years; farmer; res. North Bloomfield, Ct. No issue.

8.—**163 Lucius**, b. Aug. 8, 1827, T.; m. Mary E. Jackson.

79 Hiram Twining, son of **35** William, b. M'ch 31, 1794, T ; d. at his son Samuel's, Fulton Co., Ill., M'ch 8, 1876; m. Lovey Peace, of Maine, Dec. 14, 1820; b. April 11, 1800, d. April 12, 1850. Moved with his uncle, **38** Lewis Twining, to Licking Co., Ohio; drove the ox team for him over the mountains. In 1826 bought a farm one-half mile east of where Alexandria, Ohio, now stands, and remained there a farmer to 1864. Was a man of excellent character, intelligent, industrious and never heard to use a profane word; member of M. E. Church.

ISSUE:

1.—**Philena**, b. Oct. 15, 1821; m. Richard Stewart, M'ch 11, 1838; res. Lewistown, Ill. *Issue:*
   1. Austin W.; res. Jacksonville, Ill.
   2. James M.; served three years in the Rebellion, 121st Ohio Reg.; res. Lewistown, Ill.
   3. Annie E.; res. Cambridge, Neb.

2.—**Mary P.**, b. Dec. 17, 1823; m. Obadiah C. Houghton, Feb. 9, 1840; res. Lewistown, Ill. *Issue:*
   1. Darius K.
   2. Philena.
   3. Henrietta.
   4. Hiram.
   5. Anna.
   6. Ella.
   7. Albert.

3.—**Henry N.**, b. June 26, 1827; m. Mary A. Rogers, Oct. 23, 1849; d. April 3, 1889; photographer; res. Burlington, Iowa, where he moved 1854; one child, Medora C., b. M'ch 24, 1853; d. April 14, 1855.
4.—164 **Samuel R.**, b. Jan. 30, 1831; m. Sarah E. Overstreet.
5.—**Julia A.**, b. May 16, 1835; m. Charles D. Maranville, Dec. 29, 1859; res. Alexandria, O.; two chil., Frank W. and Fred C.
6.—**Anna M.**, b. Sept. 11, 1838; m. Charles E. Smith, Nov. 15, 1866; res. West Berlin, Ohio; served in the Civil War 4 years, 32d Ohio Reg. *Issue:*
    1. Edward.
    2. Earnest.
    3. Geo. W.

80 Joseph Twining, son of 35 William,

b. M'ch 27, 1796, T.; d. Feb. 5, 1860, So. Champion, N. Y.; m. Rachel Lewis; d. Dec. 30, 1875, aged 78 yrs, 11 mo., 19 days; farmer.

ISSUE:

1.—**James Hiram**, b. Oct. 18, 1823; d. Aug. 16, 1851, South Champion; m. Abigal Waldo, yet living (1888). *Issue:*
    1. Hiram M., b. May 29, 1848; d. Aug. 14, 1851.
    2. Charles W., b. Dec. 7, 1849; d. Aug. 14, 1851. These children were buried in one casket and their father was buried at the same time (of bloody flux).
    3. Mary A., d. infant.
2.—**Eliza A.**, b. Feb. 4, 1825; m. Mortimer Waldo, who d. 1887; res. East Watertown, N.Y.; M.E. Church. *Issue:*
    1. James H., b. Oct. 21, 1847; d. inf.
    2. Ellen E., b. Oct. 4, 1752; m. James Hodge; res. Copenhagen, N. Y.
    3. Charles M., b. Oct. 29, 1861; d. Aug. 24, 1887.
3.—**Henry M.**, b. April 19, 1828; d. Dec. 21, 1846.
4.—**Mary F.**, b. July 2, 1831; d. Nov. 7, 1839.

81 Barnabas Twining, son of 36 Eleazer,

b. M'ch 1, 1800; d. Feb. 11, 1831; m. Harriett Phelps

about 1824, dau. of Elijah, of Otis; lived in Berkshire Co., Mass.

ISSUE:

1.—**Eleazer,** b. M'ch 22, 1825; m. and had five or six chil.; "went West," but efforts to trace him have been ineffectual.
2.—**Cordelia,** b. May 22, 1827; m. Nov. 27, 1850, Solomon Clark, New Boston, Mass., who d. Sept. 2, 1882; wid.; res. Westfield, Mass. *Issue:*
    1. Ella C., b. Nov. 10, 1851; m. Jno. Mills, 1870; res. N. Haven, Ct.
    2. Catie H., b. April 19, 1855; m. Silas Ripley, 1874; res. E. Granville, Mass.
    3. Gertrude E., b. Aug. 16, 1859; d. April 22, 1878, Westfield, Mass.
    4. Burton S., b. July 16, 1861; m. Fannie Pomeroy; res. Westfield.
3.—**Ruth,** b. Oct. 8, 1830; d. Jan. 15, 1880; m. Luman Jerome, of Bristol, Ct., May 6, 1849; d. M'ch 10, 1872. *Issue:*
    1. Adelia, d. young.
    2. Adelaide, d. young.
    3. Anna, b. Dec. 10, 1859; m. Geo. W. Hamlin, Plainville, Ct.

82 Philander F. Twining, son of **37** Judah, b. May 6, 1809; d. Nov. 13, 1877; m. Sarah A., dau. of Jonathan and Abigail (Boise) Shephard, Oct. 19, 1831; d. Oct. 8, 1885, aged 77 yrs. A man eminent for piety; 30 years an official and deacon in the Tolland Cong. Church; also held various township offices. Six years prior to his death he resided at New Boston, Mass.

ISSUE:

1.—**165 Nelson B.,** b. Nov. 16, 1832; m. M. E. Webb.
2.—**Lewis T.,** b. Aug. 4, 1836; m. Hannah A. Webb (sister of above), June 7, 1865; she d. April 15, 1877; he d. Nov. 27, 1877; farmer, of Sandisfield; no issue.
3.—**166 Homer P.,** b. Nov. 9, 1839; m. Mary B. White.

4.—**Lois Etta**, b. Jan. 22, 1844; d. Jan. 27, 1860; educated at Claverack, N. Y. Sem. It is said all bearing the name "Lois" were beautiful women.

83 Merrick Twining, son of **38** Lewis, b. July 13, 1807, T.; m. *first*, Corintha Clark, Jan. 22, 1829; she d. Sept. 29, 1884; m. *secondly*, Mrs. Pierson, of Newark, Ohio, Dec., 1886; farmer; Cong.; res. Granville, Ohio.

ISSUE:

1.—**Henrietta J.**, b. Sept. 14, 1831, Berlin, O.; m. *first*, Sept. 14, 1848, Lewis Jones, b. Feb. 14, 1819, Cardiganshire, Wales, d. M'ch 23, '64; m. *secondly*, Feb. 16, 1869, Morgan Williams, b. Oct. 14, 1833, Utica, N. Y.; farmer; elder Presbyterian Ch.; res. Granville. *Issue:*

    1. De Esting W., b. July 29, 1849; hardware merchant; res. Granville.
    2. Otto S., b. April 14, 1851, m. Georgiana Williams; liveryman; res. G.
    3. Alma M., b. June 26, 1857; m. Ed. E. Tight, Nov. 30, 1887; res. G.
    4. Howard L, b. May 4, 1871 (twin).
    5. George L., b. May 4, 1871 (twin).

2.—**167 Lewis S.**, b. April 6, 1833; m. Philena C. Moore.
3.—**168 Edward W.**, b. May 8, 1836; m. Matilda Hughson.
4.—**Harriett O.**, b. Dec. 8, 1838.
5—**169 Henry L.**, b. Sept. 16, 1841; m. Annie Moore.
6—**Almira R.**, b. Sept. 10, 1846; m. Ben. S. Marshall, Oct. 18, 1865, b. Jan. 24, 1843, Knox, O.; carpenter; res. Newark, Ohio. *Issue:*

    1. Charles M., b. Dec. 7, 1866; d. Aug. 27, 1869.
    2. Birdena, b. Sept. 15, 1872, Otho, Iowa.
    3. William H., b. Nov. 3, 1878, Hartford, O.
    4. Harry G., b. Nov. 13, 1884, Girard, Kan.

7.—**170 Nelson L.**, b. Oct. 10, 1850; m. Caroline Hall.
8.—**Charles D.**, b. June 21, 1854; m. Elma Williams, Oct. 3, 1876, b. July 4, 1859; farmer; res. Hillier, Knox Co., Ohio. *Issue:*

    1. Hattie B., b. Oct. 13, 1877.
    2. Minnie, b. July 31, 1883.

9.—**Gratia M.**, b. July 22, 1860; m. Nov. 23, 1884, Dent Barrick, b. 1861; blacksmith; one child, William B., b. July, 1886.

84 Edward W. Twining, son of **38** Lewis, b. Oct. 5, 1814. Took a literary course at Ohio University and theological course at Lane Seminary during Lyman Beecher's connection therewith; entered Methodist ministry 1840 in Iowa, where he continued forty years; res. Corning, Iowa; m. *first*, Adeline Weed, Feb. 3, 1840, d. Jan. 3, 1848, Washington, Iowa; m. *secondly*, P. B. Ashley, Aug. 28, 1849.

ISSUE BY FIRST WIFE:

1.—**Jennette**, d. inf.
2.—**Almima**, d. inf.
3.—**Fenimore**, d. inf.
4.—**171 Edward T.**, b. Aug. 5, 1844; m. Florence Conger.
5.—**172 Lauriston**, b. June 7, 1845; m. Laura Botkin.

ISSUE BY SECOND WIFE:

6.—**173 Jessie L.**, b. Aug. 5, 1850; m. Flora D. Rowley.

85 Jonathan Twining, son of **39** Nathan; b. Nov. 9, 1790, Alstead, N. H., d. April, 1864, Brattleboro, Vt., hospital; m. Jan. 9, 1817, Eliza Ann, dau. of John and Eliz. Fessenden, b. in Townsend, Mass., 1802; d. Green Garden, Ill., at the home of her son, Hiram, June, 1872.

Removed to town of Shrewsbury, Vt., about 1835. He was a lay preacher of the Christian denomination; was quite gifted as an extempore speaker. It is said that "some agreeable reminiscences exist of him." By trade a cooper and shoemaker.

Rev. Edward W. Twining

THE TWINING FAMILY. 121

ISSUE:

1.—**174 Hiram**, b. June 9, 1819; m. Betsey Needham.
2.—**Lucina B.**, b. May 20, 1821, Gilsum, N. H.; m. Nov. 12, 1846, Oliver Woods, b. Petersboro, N. H., Feb. 15, 1811, son of Nehemiah and Jerusha (Stevens) Woods; res. Manchester, N. H.
3.—**Bemsley Lord**, b. Sept. 29, 1822; farmer; unm.; res. Shrewsbury.
4.—**Paschal W.**, b. July 25, 1825, Gilsum; d. Aug. 22, 1852; unm.
5.—**Merinda F.**, b. July 12, 1828; m. Samuel C., son of Brooks and Sally (Clark) Hudson, Dec. 1, 1846; she d. Nov. 27, 1856, leaving one child, a dau. He m. again; res. Gilsum, N. H.
6.—**Sarah E.**, b. April 1, 1835, Shrewsbury; m. 1858, Abel Spaulding; res. Townsend Centre, Mass.; one child, Alice, b. Oct. 1865.
7.—**Elmira F.**, b. May 12, 1837, S.; m. Feb. 18. 1856, Alpheus Smith; res. Forest Home, Franklin Co., Kan. *Issue:*
    1. Samuel A., b. in Ill., 1858.
    2. Clara, b. in Ill., 1861.

86 Ebenezer Twining, son of **40** Barnabas, b. April 4, 1801, Orleans, Mass; d. Oct. 3, 1877, Swampscott, Mass., where he removed M'ch, 1843; fisherman; Universalist; m. Meribah Small (dau. of Zacheriah, of South Orleans); d. Nov. 25, 1877, S., aged 74 years; Baptist.

ISSUE:

1.—**Rebecca**, b. Sept. 13, 1824; m. *first*, Ensign Eldridge, April 8, 1840, who d. Sept. 8, 1853, So. Chatham, Mass.; m. *secondly*, Luther Eldridge, cousin of first husband, May 10, 1862, who d. Feb. 11, 1879, So. Chatham, where wid. res. 1889; "Come Outers." *Issue:*
    1. Clement, b. Oct. 1, 1845; m. Susan M. Eldridge, 1870, So. C.; book pub.; res. Battle Creek, Michigan; Seventh Day Advent.

  2. Ensign A., b. June 5, 1848; m. *first*, Ellen M. Crowell, 1867, d. 1877; m. *secondly*, O. E. Nason, 1886; merchant; res. Lynn, Mass.; M. E. Ch.
  3. Sybrina T., b. July 10, 1850; d. Nov. 6, 1851.
  4. Alonzo, b. July 25, 1852; m. Ella Nickerson, of E. Harwich, Mass., 1876; she d. 1886; mariner; res. So. Chatham, Mass.
  5. Henry W., b. July 17, 1863; m. Mary S. Hunt, of Lynn, Mass. 1886; poultry business; res. So. Chatham, Mass.

2.—**Sabrina**, b. Jan. 15, 1827; d. 1848.

3.—**Barnabas**, b. Sept. 19, 1830; d. young.

4.—**Malvina W.**, b. Sept. 17, 1831; d. Dec. 1857; m. John Tuttle, of Dedham, Mass. *Issue:*

  1. Amanda W., b. about 1854; d. inf.
  2. Willie, b. 1857; d. Aug. 1858.

5.—**Meribah**, changed to *Mary Francis* (twin), b. May 4, 1836; m. Nov. 27, 1856, Edward Marsh, Jr.; fisherman; M. E. Church; res. Swampscott, Mass. *Issue:*

  1. George H., b. Sept. 29, 1857; d. Sept. 30, 1858.
  2. Ida M., b. Aug. 16, 1859; m. M'ch 5, 1879, Charles Parrott; chil.
  3. Edelena, b. Feb. 8, 1862; m. Walter F. Gage, of Wilton, N. H., Nov. 27, '89; glass blower; res. Cambridgeport, Mass.
  4. Ebenezer H., b. July 28, 1865; d Feb. 16, 1866.
  5. Edward F., b. Aug. 28, 1868; d. inf.
  6. Maud L., b. April 13, 1873.
  7. Annie L., b. Aug. 16, 1878; d. inf.

6.—**Ebenezer** (twin), b. May 4, 1836; d. Sept. 1859; drowned; m. Mary Pierce; res. Swampscott; one child, *Willie*, b. Dec. 1859; d. Aug. 1860.

7.—**Elizabeth**, b. Nov. 13, 1838; m. W. Henry Thomas, M'ch, 1859; fisherman; M. E. Ch.; res. Swampscott. *Issue:*

  1. Eliza A., b. M'ch 17, 1860; m. Edward Foye; baggage master; res. Lynn, Mass.
  2. Henrietta, b. Dec. 1863; d. June, 1876.
  3. Charles W., b. Nov. 1866; d. 1876.
  4. Mamie F., b. Nov. 1868; m. Nov. 27, 1889, Fred. W. Newhall; merchant; res. Lynn, Mass.
  5. Lettie L., b. about 1871; d. 1872.
  6. Walter A., b. Aug. 3, 1874.
  7. Nellie L., b. Sept. 4, 1877.

8.—**Barnabas**, b. Sept. 24, 1841; d. inf.

87 Addison Twining, son of **41** Abner, b. June 24, 1810, Frankfort, Maine, d. April 3, 1882, F.; m. Jan. 6, 1849, Emeline, dau. of Henry Colson, b. April 12, 1831. Lived on the old homestead at F., where all his children were born; wid. res. in Winterport, Me.

ISSUE:

1.—**Albert T.**, b. Oct. 17, 1849; m. Jennie Cole, of Winterport, who d. Feb. 1887. Served in the 20th Maine Reg. in the Rebellion; farmer; res. F.; no issue.
2.—**Alice O.**, b. Nov. 5, 1850, d. M'ch 9, 1853.
3.—**Violet A.**, b. Feb. 7, 1852; m. L. B. Grindle, of Bluehill, Me., Dec. 31, 1872; granite cutter; no chil.; res. Bluehill.
4.—**Addison, Jr.**, b. Nov. 2, 1853; went to Woburn, Mass., when eighteen years old, where he is engaged in farming; m. Mazzie Graham. *Issue:*
    Mable, Emma and Violet, and an inf. dau.
5.—**Arthur F.**, b. Feb. 26, 1856; d. Nov. 22, 1858.
6.—**Nathan F.**, b. June 26, 1861. Lived at F. until 20 years old, when he went to Bluehill, Me., and learned the granite cutters trade; res. Belfast, Me.

88 Harrison Twining, son of **41** Abner; b. Nov. 14, 1814, Frankfort, Me., where he resided until 1885, when he moved to Au Gres, Michigan; fell from a hammock, causing paralysis, from which he d. Nov. 5, 1887. He was a prominent Universalist minister, a brilliant conversationalist, a man well read and of sterling worth. Married, *first*, Olive Higgins, Oct. 20, 1842, b. Nov. 13, 1824, d. Sept. 20, 1845; two chil ; d. inf.; m. *secondly*, Bethiah Higgins (sister of first wife), June 10, 1846, d. Oct. 16, 1846; m. *thirdly*, Mary Jane Cole (cousin to first two), April 27, 1847, b. May 22,

1829; she res. Au Gres, Mich., and on Thanksgiving (1889) gave a "real old down East styled dinner."

ISSUE BY THIRD WIFE:

1. —**Howard**, d. inf.
2. —**Olive Rosilla**, b. M'ch 1, 1848; m. F. A. Gilkey, of Houlton, Me.; res. Newell, Iowa; one child, Maud, b. about 1876.
3. —**Howard**, b. Dec. 1, 1849; d. M'ch 3, 1850.
4. —**Sarah**, b. Feb. 22, 1851; m. E. G. Cole, Feb. 27, 1873, b. Hampden, Me., Jan. 25, 1851; res. Au Gres, Mich.; no children.
5. —**Lizette**, b. Oct. 16, 1853; d. Aug. 9, 1869.
6. —**Harrison**, b. Oct. 18, 1855; d. Jan. 15, 1873.
7. —**Herbert**, b. Jan. 15, 1857; d. M'ch 8, 1883.
8. —**Elvira Cobb**, b. M'ch 5, 1860; m. Nov. 30, 1881, James Grimore, lumberman and merchant; res. Au Gres, Mich. *Issue:*
    1. Gail G., b. Feb. 26, 1883.
    2. Pearle Twining, b. Sept. 24, 1884.
9. —**Fred L.**, b. Sept. 30, 1865; m. Lillie Hill Sept. 15, 1888; bookkeeper; res. Au Gres, Mich.
10. —**Willie S.**, b. Jan. 15, 1868; d. Jan. 8, 1881.
11. —**Hannah**, d. inf.

89 Jonathan Twining, son of **42** Jonathan; b May 13, 1797, O., d. Oct. 10, 1846, New Orleans, La., of yellow fever. He was captain of a little brig which sailed to and from "Perloma," and also sailed from Boston to foreign parts and d. on his last voyage; m. *first*, Sukey Linnell, April 2, 1819; baptized 1799; d. March 31, 1843, aged 43 yrs., O.; m. *secondly*, Sarah Cook, of Boston, Mass.

ISSUE:

1. —**Joseph**, b. Oct. 5, 1820; d. Sept. 21, 1821.
2. —**Lucy**, b. April 3, 1823; d. Oct. 27, 1873, Chicago, Ill.; m.

Ed. C. Cooledge, of Hillsborough, N. H. Left one son, who m. Carrie Haight, of Aurora, Ill.; res. Chicago.

3.—**Jonathan**, b. April 29, 1825. Went to sea in the ship Alice Gray, of Boston, Freeman S. Nickerson, of Boston, master, on her passage from Philadelphia to Londonderry, in April, 1847. She had a cargo for the famine stricken Irish, but was never heard from after leaving port, and it is certain the vessel and all on board found watery graves in the Atlantic. Jonathan was one of the officers of the ship.

4.—**Hannah J.**, b. Jan. 27, 1828; d. Sept. 10, 1847, E. Boston, unm.

5.—**Joseph R.**, b. Sept. 18, 1830; d. of consumption at the Marine Hospital, Boston, before 1855; unm.

6.—**Sukey** or Susan, b. M'ch 6, 1833; d. Lowell, Mass.; m. Norman Cobb, of Johnsbury, Vt., who d. in Cal.; left one child, Wallace W., who m. Eliz. Rose; he d. Nov. 1883.

7.—**Tamzen**, b. M'ch 9, 1835; m. Feb. 1857, Nathaniel K. Holland, of Newburyport; repair shop; res. Salem, Mass. One child, Geo. H., b. 1858; m. Eliz. Bosworth, of Stoneham, Mass.

8.—**175 George F.**, b. Jan. 28, 1838; m. Annie Whittier.

EIGHTH, NINTH AND TENTH GENERATIONS.

90 Stephen B. Twining, son of **44** Charles, b. Jan. 19, 1844; m. Letetia Warner, Jan. 17, 1866; Friend; engaged in stone quarries with his brother, **91** Ed. W.; res. Yardley, Pa.

ISSUE:

1.—Sarah W., b. March 4, 1867.
2.—Elizabeth, b. M'ch 19, 1870.

91 Edward W. Twining, son of **44** Charles, b. M'ch 4, 1846; m. Mary S. Walker, Jan. 14, 1878; Friend; res. Yardley, Pa.

ISSUE:

1.—Charles T., b. Dec. 30, 1878; d. Jan. 30, 1879.
2.—Stephen Baldwin, b. Dec. 29, 1879.
3.—Ernest W., b. Dec. 21, 1883; d. Feb. 17, 1885.

92 Aaron Twining, son of **45** John, b. M'ch 31, 1827; d. Aug, 25, 1862, Oldtown, Ark., during service in late war, Co. C, 11th Wis. Vol.; res. Medina, Wis. "A generous and sympathetic man;" farmer; M. E. Ch ; m. Mary Lyons, Dec. 31, 1853, b. May 31, 1836. She again m. Jerry Folsom, Sept. 13, 1868, and moved to Alexandria, Dak., 1884, by whom she has five children.

THE TWINING FAMILY. 127

ISSUE BY FIRST WIFE:

1.—William H., b. Dec. 15, 1854, Waterloo, Wis.; m. Mary A. Mitchell, June 9, 1881, b. June 30, 1859; moved to Alexandria, Dak., 1880. Owns 480 acres of Dakota "dirt." *Issue:*
    1. Geo. M., b. M'ch 2, 1882; d. Nov. 26, 1884.
    2. Bell, b. Dec. 20, 1886.
    3. Leon Robert, b. Sept. 9, 1889.

2.—James Carey, b. April 13, 1856; res. Minor, Dak.; owns 160 acre farm; unm.

3.—Emma J., b. April 12, 1859; m. Delacy Betts, Nov. 29, 1883; b. April 10, 1858; res. Sioux Falls, Dak. *Issue:*
    1. Julian, b. Sept. 18, 1884; d. Feb. 13, 1885.
    2. Mattie E., b. Feb. 9, 1886; d. Feb. 19, 1887.
    3. A son, b. Feb. 1888.

93 Nathan C. Twining, son of **45** John; b. Sept. 27, 1834; m. *first*, Phebe Ann Barber, Nov. 18, 1861, b. June 16, 1838, Hopkinton, R. I., d. Jan. 16, 1866, Edgerton, Rock Co., Wis.; a Christian lady of refined manners and finished education; m. *secondly*, Mary Jane Rennie, M'ch 23, 1868.

From the Wisconsin Ed. Record the following extracts are obtained relating to our subject:

"Among the most prominent of Wisconsin's educators and teachers is Prof. Nathan C. Twining, of Monroe, Wis. Having received a thorough academic, as well as collegiate education, he stands second to none in his profession. He has wielded a great influence in the State in all the public school matters, engrafting his ideas not only on the minds of teachers, but has been influential in giving shape to *their* ideas, thus making himself felt throughout the length and breadth of the State.

"He has been regarded for years as the best mathematical teacher in the State, having for eight years held the chair of Math. in Milton Coll. * * * * * * * He has given his life energies to schools of the State, having spent his time from Sept, 1873, to June, 1886, in the public schools of Monroe, Wis. He raised them to the highest rank, his school being regarded at the State University as the very best school in the State for fitting young men and women to enter upon an advanced course of university education.

"Having spent years in study of Latin and Greek, and his vacations in extensive travel, he has long held the foremost place also as a thorough instructor in classical languages, history and sciences.

"During the war of the States he held the position of captain in the 40th Reg. Wis. Vol. Inft., Co. C, being also a member of the 13th Vol.

"In person tall, finely developed, health robust, ambition that knoweth no defeat. Reared a Quaker, he strictly adheres to their strict code of morals. Absolutely free from the use of all narcotics, he is a firm and unflinching advocate of individual temperance prohibition.

"As an author and writer he stands amongst the foremost of his associates. His published papers on astronomy, language and pedagogics, science and current topics have been well received and placed in the educational archives of the State."

In 1887 he took charge of the Riverside, Cal., schools, where he also owns a vineyard and orange grove. (See portrait page 86)

ISSUE BY FIRST WIFE:

1.—**Harry La Verne**, b. M'ch 8, 1863, Milton, Wis.; express office clerk, Georgetown, Col., but now assistant in Riverside, Cal., schools.

2.—**Clarence W.**, b. April 27, 1864, M.; ass't bank cashier; res. Monroe, Wis.; a first-class penman; m. Sept. 30, 1886, Mazie V. Barber, b. June 17, 1867, dau. of Joseph C. Barber and Louisa A. Rittenhouse. *Issue:*

    1. Joseph Laverne, b. Sept. 17, 1887.
    2. Phebe L., b. Sept. 22, 1889.

ISSUE BY SECOND WIFE:

3.—**Nathan Crook, Jr.**, b. Jan. 17, 1869, Boscobel, Wis. Graduated with honors at Monroe High School, 1884; 1885 at Valparaiso (Ind.) Normal School, where he became proficient in stenography and typewriting. Entered U. S. Naval Academy at Annapolis, Md., Sept. 1885, where he stood number six in a class of eighty-six. Grad. June, 1889, number four in a class of thirty-eight. At once entered a two years' cruise on board the U. S. steamer "Chicago." A young man of fine form, robust, teetotaler and master of five languages. "Successus meritos coronet."

94 Henry Harrison Twining, son of 45 John,

b. Feb. 11, 1841; m. Hattie E. Miller, Nov. 14, 1865,

CADET N. C. TWINING, JR.
U. S. Steel Cruiser Chicago, U. S. N., 1890.

b. Nov. 6, 1845, Dane Co., Wis. Lived in Grundy, Butler and Mitchell counties, Iowa, to Oct. 1886, when he settled at Hiram, Cleburne Co., Arkansas; farmer; Baptist and Sunday-school worker. Served four years and twenty-six days in Co. C, 11th Wis. V. V. I. R Fired first shot on Vicksburg. For bravery and daring was twice promoted.

ISSUE:

1.—**Addie Lorena**, b. July 24, 1868, Grundy Co., Iowa.
2.—**Burton Miller**, b. Oct. 25, 1871, Grundy Co., Iowa.
3.—**Ralph Waldo**, b. April 14, 1873, Butler Co., Iowa.
4.—**Earnest Centennial**, b. Sept. 21, 1877, Butler Co., Iowa.
5.—**Earl**, b. Oct. 29, 1884, Osage, Iowa; d. Oct. 27, 1887, Hiram, Ark.

95 Peter Slater Twining, son of **45** John, b. Feb. 27, 1844; m. Cornelia Z. Cooper, Feb. 25, 1863, b. Feb. 13, 1844, Wilton, N. Y. Sold his farm, a part of the old homestead at Waterloo, Wis., 1886, on which he had resided 42 years; res. W.

ISSUE:

1.—**Lionall A.**, b. Jan. 5, 1865.
2.—**Lillie M.**, b. Jan. 20, 1869; d. Feb. 6, 1870.
3.—**Arthur F.**, b. Aug. 31, 1872.
4.—**Rose M.**, b. Dec. 3, 1874.
5.—**Paul E.** (twin), b. May 4, 1877.
6.—**Perry J.** (twin), b. May 4, 1877.
7.—**Ray C.**, b. Oct. 8, 1878.

96 Chapin Twining, son of **46** Charles, b Jan. 16, 1817; m. wid. Kee (maiden name Levisa Whitaker), July 15, 1863, b. May 7, 1837; farmer; res. Potosi, Grant Co., Wis.

THE TWINING FAMILY.

ISSUE:

1.—**Melissa**, b. June 24, 1864; m. March 24, 1887, Frank Schwartz.
2.—**John H.**, b. Dec. 1, 1866; d. Oct. 1, 1881.
3.—**Amanda A.**, b. July 10, 1869.
4.—**George A.**, b. July 21, 1872.
5.—**Addison L.**, b. June 20, 1875; d. Oct. 7, 1881.

97 Dewitt C. Twining, son of **47** Thomas, b Sept. 23, 1824, North Boston, Erie Co., N. Y.; m *first*, (second cousin) Susannah G. Hambleton, dau. of Moses and Ann (Galbreath) Hambleton,* 1849, Lagro, Ind., b. New Lisbon, Ohio, Oct. 25, 1827, d. June 5, 1884, Roanoke, Ind. A truly noble Christian mother and companion, whose life was well spent, whose memory is sacredly revered by many, and especially her children; a birthright Quaker, but d. member of M. E. Ch.; m. *secondly*, Mary E. Webster, Jan. 20, 1886; divorced 1889. The major portion of his life has been spent in Huntington Co., Ind.; raised a farmer; Apiarist. In 1853 went by "Overland Route" to Cal., where he remained one year; served in the Union army, 1865, Co. K, 13th Ind. Vol.; res. Roanoke, Ind.

* Moses Hambleton, b. Dec. 26, 1796, Bucks Co., Pa.; d. Nov. 5, 1856, near Fort Wayne, Ind.; m. about 1823, Ann, dau. of James and Susannah Galbreath, from Guilford Co., N. C., of Scotch stock. Moses moved from Erie Co., N. Y., to New Lisbon, Columbiana Co., Ohio, 1819; resided there, a miller, till 1834, when he settled with his family in Camden, Jay Co., Ind Was one of the first settlers in that county and had to go forty miles to mill. In 1842 removed to "Quaker Settlement," near Huntington, Ind.; engaged in farming; was a good story-teller; much devoted to hunting. Quaker, as were his and wife's ancestors.

He was a son of Aaron and Hannah (Kester) Hambleton, of Bucks Co., Pa., who moved to Upper Canada, 1809; she d. 1815 and he then settled in Boston, Erie Co., N. Y.; m. *secondly*, Naby Stone and d. May 3 1829.

For a record of the Hambletons we are indebted to C. J. Hambleton, of Chicago, Ill., who compiled and pub. the Hambleton Genealogy, 1887.

ISSUE:

1.—**Estella G.**, b. Oct. 8, 1849; unm.; res. Roanoke, Ind.
2.—**Thos. J.**, b. Jan. 30, 1851, Huntington, Indiana, in "the old log hut" on the Wabash. Educated at the Roanoke Seminary. Spent his boyhood days mostly on the farm. After reaching majority engaged in the wagon business, subsequently loans, real estate and insurance. Prior to 1882, res. Roanoke, Ind., since said date, Sidney, Kosciusko county, Indiana. A teetotaler; member Prohibition party; Baptist, but believes in conditional immortality, the pre-millenial coming of Christ and his reign till the final extinction of evil; slender build, above the average height, nervous-sanguine temperament, robust health, deeply interested in reforms of the day, and thinks life is worth living. Compiler of the Twining Genealogy. Owns a section of land near Pierre, South Dakota; m. May 10, 1876, Margaret A., dau. of Wm. and Martha (Norris) Cordill, of Whitley Co., Ind.; b. Sept. 15, 1851, d. of consumption, Dec. 1, 1887, Sidney, Ind. A very earnest, kind and devoted wife and mother; Baptist. *Issue:* (See frontispiece.)
    1. Leonora, b. Aug. 26, 1877, Roanoke, Ind.
    2. Arthur, b. M'ch 1, 1882, Roanoke, Ind.
    3. Earl, b. May 18, 1884, Sidney, Ind.
3.—**Moses D.**, b. Aug. 8, 1853; d. inf.
4.—**Rosetta O.**, b. April 23, 1855, Erie Co., N. Y.; m. W. E. Callison, May 10, '76, So. Whitley, Ind., b. Jan. 30, 1854; son of Wm. W. and Julia A. Callison; school teacher and agricultural dealer; res. Huntington, Ind.; no issue.
5.—**Rachel A.**, b. Jan. 26, 1858; m. John H. Thorn, 1880, b. 1862; U. B. Ch.; res. St. Joseph, Mo. *Issue:*
    1. Orville A., b. M'ch 29, 1881.
    2. Birtie, b. M'ch 26, 1885.
    3. Wilber R., b. July 9, 1887.
    4. —— —— (twin), b. July 9, 1887; d.
6.—**Franklin L.**, b. July 3, 1860; d. 1861, Roanoke.
7.—**Sarah M.**, b. Sept. 30, 1862; m. Jacob D. Lininger, Nov. 15, 1883; b. Sept. 19, 1860; res. Huntington, Ind.; dry goods business; Evangelical Ch.; child, Earl, b. July 20, 1885, H.
8.—**Bertha E.**, b. Jan. 13, 1865; d. inf.
9.—**Leonard M.**, b. Jan. 19, 1868; d. inf.

98 Lewis Twining, son of **47** Thomas, b. M'ch 24, 1830, d. Oct. 24, 1871, Topeka, Kansas. Was engaged in farming, North Collins, Erie Co., N. Y., till about 1860. Became a prominent Apiarist, traveling much in the Wabash Valley and the West. An intelligent but eccentric man; m. *first*, Mary E. Sherman, Oct. 23, 1853, b. Sept. 12, 1834, Fall River, Mass; a well-informed woman of many excellent qualities; grad. C. L. S. C.; Friends Society; res. Lawton Station, Erie Co., N. Y.

ISSUE:

1.—**Edward Ellis**, b. Dec. 5, 1854; prominent farmer and school teacher; Republican; a man of good abilities; m. Chloe Smith; she was educated Fredonia Normal School; res. Lawton Station, N. Y.; no issue.
2.—**Emma A.**, b. June 6, 1856; accomplished teacher, typewriter and stenographer; grad. C. L. S. C.; Friend; res. L. Station; single.

99 Charles Twining, son of **48** Thomas, b. Aug. 23, 1822, Groton, N. Y. Educated at Groton Academy. In 1847 went to Lancaster Co., Pa., to teach and remained there till the fall of 1856, in which year he organized the Warren, Pa., Borough Union School, which he conducted as Prin. to 1863; thence to Waterford, Erie Co., Pa.; engaged in other business till 1870; 1870 to 1878, Supt. of Union City, Pa., schools; 1878 to 1884, Supt. Erie Co., Pa., Common schools. Moved to Sumner Co., Kansas, 1885; m. *first*, Mary Stanton, July 30, 1850, Lancaster, Pa., b. Feb. 29, 1829; d. Aug. 27, 1873, of consumption; m.

Prof. CHARLES TWINING.
(See Page 133.)

secondly, Jennie E. Terry, July 29, 1875, b. Wayne Tp., Erie Co., Pa., July 21, 1845; members M. E. Church; res. Argona, Kan., where he owns a farm. (See engraving.)

ISSUE:

1.—**Sybilla Heithue**, b. May 5, 1851; m. Geo. W. H. Reed, of Union City, Sept. 10, 1872; insurance agent. *Issue:*
   1. Mary Lila, b. Jan. 19, 1876.
   2. Bessie, b. April 4, 1879.
   3. Georgia A., b. Oct. 16, 1882.
2.—**Charles A.**, b. April 4, 1854, L.; m. Jan. 1, 1885, May Reynolds; bookkeeper; res. Corry, Pa.; one child, Beatrice Almira, b. Jan. 5, 1888.
3.—**William S.**, b. Feb. 20, 1865, near Titusville, Pa.; 1886, attending Allegheny Coll. Pa.
4.—**Walter Clare**, b. Feb. 14, 1868, Waterford; res. Argona, Kan.; grad. Wichita, Kan., Business Coll., 1888.

100 James Twining, son of **49** John, b. Aug. 10, 1817; m. *first*, M'ch 15, 1841, Rebecca Howard; m. *secondly*, Frances Benedict; m. *thirdly*, Alice Crocker; farmer; res. Seneca Falls, N. Y.

ISSUE BY FIRST WIFE:

1.—**Lena**, d.
2.—**Lula M.**, d.

ISSUE BY THIRD WIFE:

3.—**James H.**, d. inf.
4.—**Freddy**, b. April, 1883.

101 Thomas Twining, son of **49** John, b. Aug. 11, 1819, d. Sept. 19, 1855, Union, N. Y.; m. Lucy Heald Balch, Jan. 14, 1849; teacher and merchant; Presbyterian.

ISSUE:

1.—**Thomas Dick**, b. Oct. 22, 1849; m. Dec. 22, 1869, Dorinda Cogswill, b. April 30, 1847, Vestal, N. Y.; res. Binghampton, N. Y. *Issue:*
   1. Guy E., b. April 7, 1871.
   2. Mertie L., b. M'ch 7, 1873.
2.—**Mary Stanley**, b. Aug. 22, 1851; telegraphy, bookkeeper and teacher in Elmira (N. Y.) Business Coll. Now (1890), stenographer, Buffalo, N. Y.

102 William Twining, son of 49 John, b. Sept. 23, 1822; m. Nov. 4, 1850, P. R. Miner, b. Nov. 19, 1832; res. Hooper, N. Y.; farmer.

ISSUE:

1.—**Frank B.**, b. Oct. 23, 1851; m. Nov. 3, 1880; merchant; res. Hooper, N. Y.; chil.: Edith and Lewis.
2.—**Eleanor**, b. Aug. 28, 1853; d. Sept. 15, 1855.
3.—**Eugene A.**, b. Sept. 20, 1856; m. Nov. 11, 1880; one child, Helen.
4.—**Alice L.**, b. July 3, 1857.
5.—**Fred. G.**, b. May 31, 1866.

103 John A. Twining, son of 49 John, b. Oct. 16, 1827, d. Sept. 28, 1868, Union, Broome Co., N. Y.; m. Emily Roberts, May 27, 1852.

ISSUE:

1.—**Eliza**, b. Feb. 25, 1854; d. M'ch 29, 1862.
2.—**John F.**, b. May 27, 1860; m. Mary E. Lynch, at Cedar Rapids, Iowa, June 28, 1883, b. Jan. 1865; bookkeeper; res. Cedar R.

104 Charles Twining, son of 49 John, b. April 16, 1831; m. April 30, 1856, Lucy A. Gibbs, b. April 24, '38; farmer; M. E. Ch.; res. Broome Co. (Hooper), N. Y.

THE TWINING FAMILY. 137

ISSUE:

1.—**Ida**, b. Feb. 4, 1858; m. Geo. H. Baldwin, of Broome Co.; three chil.
2.—**Dorcas**, b. June 23, 1860; m. L. C. Adams, Jan. 19, 1880.
3.—**Pollie**, b. Oct. 4, 1862.
4.—**Seymour G.**, b. Jan. 22, 1865; school teacher and farmer; res. Hooper, N. Y.
5.—**Addie A.**, b. M'ch 27, 1867.

105 Philip Twining, son of **49** John, b Aug. 13, 1833; m. Dec. 31, 1855, Francis A. Councilman, b. June 14, 1835: local Methodist minister and farmer; res. Union Centre, N. Y. Two years in the great Rebellion.

ISSUE:

1.—**Ellen L.**, b. July 25, 1857; d. Oct. 28, 1874.
2.—**Florence**, b. M'ch 20, 1859; d. Nov. 26, 1874.
3.—**Nellie A.**, b. Aug. 7, 1861; d. Oct. 23, 1874.
4.—**Burtie**, b. July 29, 1870.
5.—**Burr**, b. Sept. 19, 1870; m. Nov. 1889, Miss —— Heath; res. Union.
6.—**Howe**, b. April 19, 1875.

106 Joseph N. Twining, son of **50** Samuel, b. Nov. 12, 1818; m. Ruth A. Ames, in Broome Co.; b. July 11, 1815, d. July 19, 1882, moved to Ohio after m.; carpenter and farmer; res. Camden, Lorain Co., O.

ISSUE:

1.—**Sarah J.**, b. Jan. 28, 1844; m. J. M. Hesser; blacksmith; chil.
2.—**Mary**, b. May 20, 1845; d. M'ch 1, 1848.
3.—**Elizabeth**, b. Oct. 9, 1846; d. July 5, 1878; m. Wm. Howe; d. July 18, 1882. *Issue:*
    1. Orra E., b. about 1865; m. Theo. Smith.
    2. Carrie J., b. about 1867; m. Chas. Gill.
    3. Edwin, b. about 1870.

4. Nina C , b. about 1872.
5. Mina, b. about 1873; deceased.
6. Netta, b. about 1874.
7. Willie, b. about 1876.

4.—**Rozette**, b. April 26, 1849; m. Milo Gibson; farmer.
5.—**Frank J.**, b. Jan. 23, 1851; m. Nov. 29, 1876, Emma J. Bates; farmer and butcher; res. Kipton, Lorain Co., O. *Issue:*

1. Della M., b. Aug. 29, 1877.
2. Elsie M., b. May 22, 1880.
3. Joseph N., Jr., b. July 21, 1882.
4. Glendora, b. Oct. 10, 1888.

107 Charles Alex. Twining, son of **50** Samuel, b. May 23, 1821: m. Nellie Schermerhorn, Oct. 18, 1842, by Jessie Richards, Esq., b. Oct. 8, 1824. Came from Broome county, N. Y., to Lorain Co., Ohio, 1849, with $500 to begin farming. Owns eight good farms in L.: stock raiser and real estate dealer. He and his sons all abstain from tobacco and intoxicants, and are "good sound Democrats:" Christian connection: res. Kipton, Lorain Co., O.

ISSUE:

1.—**Sarah A.**, b. Feb. 11, 1844; m. Le Grand Gibson, April 1, 1863, b. May 11, 1837; res. West Clarksfield, Huron Co., Ohio. *Issue:*

1. Charles D , b. Jan. 27, 1864.
2. Florence O., b. Feb. 23, 1866.
3. Clarence A., b. Jan. 31, 1868.
4. Leon F., b. Feb. 9, 1870.
5. Cora B , b. Feb. 2, 1872.
6. Effa L., b. June 3, 1874.
7. Albert L., b. June 20, 1876.
8. Bertha L., b. Sept. 25, 1878.
9. Effie M., b. Sept. 10, 1880.
10. Mable, b. Oct. 25, 1882.
11. Lottie A., b. Aug. 19, 1885.

2.—**Charles H.**, b. April 17, 1846; d. Dec. 10, 1846.

3.—**William Tracy**, b. Sept. 5, 1847; m. Drucella A. Buckley, Dec. 6, 1865, b. Oct. 25, 1846; farmer; res. Henrietta Tp., Lorain Co., O. *Issue:*

    1. Estella J., b. June 23, 1870.
    2. George E., b. Jan. 3, 1875.
    3. Minnie E., b. Nov. 18, 1880.

4.—**Gertrude Eliz.**, b. Sept. 4, 1849; d. April 22, 1870.
5.—**Orlando**, b. Nov. 20, 1851; d. Sept. 21, 1853.
6.—**Alvah P.**, b. May 28, 1854; m. Sarah J. Herbert, Jan. 28, 1877, b. Aug. 15, 1854; farmer; res. Henrietta Tp., Lorain Co., O. *Issue:*

    1. Charles Alex., b. Nov. 17, 1877.
    2. Nellie M., b. Dec. 21, 1879.
    3. Sarah A., b. June 7, 1882.
    4. Isaac L., b. M'ch 11, 1884.
    5. Fred. A., b. June 28, 1886.
    6. Amy, b. April 17, 1889.

7.—**Floyd O.**, b. Sept. 16, 1856; m. Nettie Goss, Oct. 10, 1878, b. Feb. 24, 1859; farmer; Lorain Co., O. *Issue:*

    1. Josie M., b. Oct. 30, 1883.
    2. Maud E., b. March 15, 1889.

8.—**Virgil L.**, b. M'ch 11, 1859; m. Adell H. Fox, Nov. 23, '80, b. Feb. 4, 1863; res. Lorain Co., O. *Issue:*

    1. Herbert A., b. Feb. 15, 1881.
    2. Olive B., b. Sept. 8, 1882.
    3. Leland R., b. July 13, 1885.

9.—**Perry E.**, b. Feb. 21, 1853; m. Feb. 21, 1883, Mary L. Beecher, b. May 20, 1864; res. Camden Tp., Lorain Co., Ohio. One child, Rosamond E., b. Feb. 20, 1884.
10.—**Fred. A.**, b. June 30, 1865; a young man of promise; teacher and penman.

108 David Meeker Twining, son of 51

Benjamin, b. July 25, 1832; m. Phebe A. Evans, M'ch 27, 1861. Enlisted in army, Aug. 1862, and served till close of war. From exposure in the service he contracted heart disease; farmer and mechanic; res. Soldiers' Grove, Crawford Co., Wis.

ISSUE:

1.—**Etta**, b. July 28, 1866, d. July 20, 1889; teacher; m. Sept. 13, 1888, Edward M. Calkins, merchant; res. Bagley, Wisconsin. *Issue:*
    One son, Harry Twining, b. July 10, 1889.
2.—**Benjamin H.**, b. Jan. 3, 1870.

109 Joseph Twining, son of **52** Mahlon, b. Jan. 11, 1826; m. *first*, Emeline Birdsell, M'ch 20, 1847, b. Nov. 12, 1827, d. Nov. 7, 1861; m. *secondly*, Delia Dimon, Jan. 8, 1862, b. Aug. 25, 1836; farmer; res. Friendsville, Pa., where he has resided over forty years.

ISSUE BY FIRST WIFE:

1.—**Phebe A.**, b. May 20, 1850; m. Wm. Fotton, June 9, 1875, b. July 8, 1852; farmer; res. Kirkwood, N. Y. Children: *Eddie C., Frank M., Louella E., Fayette W.* and *Jessie M.*
2.—**George F.**, b. July 29, 1853; m. Oct. 27, 1880, Nella A. Meeker, b. Sept. 30, 1864; lumber business; res. Binghampton, N. Y. *Issue:*
    1. La Verne M., M'ch 16, 1882.
3.—**Clarisa A.**, b. Nov. 13, 1857; d. Jan. 6, 1881.
4.—**William J.**, b. Jan. 9, 1859.
5.—**Eliz. S.**, b. M'ch 4, 1861; m. Feb. 13, 1884, Nelson R. Bunts, b. July 28, 1859; stage driver; res. Friendville, Pa.; one child, Floyd T., b. Sept. 15, 1885.

ISSUE BY SECOND WIFE.

6.—**Fred S.**, b. Sept. 23, 1862.
7.—**Emma M.**, b. Feb. 5, 1866.
7.—**Della M.**, b. Sept. 4, 1878.

110 Chester P. Twining, son of **52** Mahlon, b. M'ch 13, 1829; m. *first*, Ann Defan (widow), from whom he was divorced after living together eight or nine years; m. *secondly*, Fanny Hayes. Lived in Marion and Carroll counties, Iowa; res. Audubon, Iowa.

ISSUE BY FIRST WIFE:

1.—**Kossuth**, d. young.
2.—**Jennie**, d. young.
3.—**Carrie.**
4.—**Desire Ann.**

ISSUE BY SECOND WIFE:

5.—**Minnie**, b. about 1869; m. about 1882, —— Rayne.
6.—**Albert E.**, b. about 1871.

111 Frederick F. Twining, son of **52** Mahlon, b. Oct. 27, 1831; m. Helen F., oldest dau. of Jno. T. Payne, Oct. 24, 1855. He was killed at battle of Lookout Mt., Nov. 24, 1863, member 137th N. Y. Vol.; Baptist; wid. res. Binghamton, N. Y.

ISSUE:

1.—**John P.**, b. M'ch 16, 1860; m. Oct. 6, 1886; res. Mulberry, Saline Co., Kan.; one child, Mary H., b. Sept. 29, 1887.

112 William F. Twining, son of **52** Mahlon, b. Sept. 27, 1834, d. April 18, 1872: m. M'ch 27, 1865, Eleanor Keyes; farmer; Baptist; res. East Maine, N.Y.

ISSUE:

1.—**Frank K.**, b. Jan. 31, 1866; farmer; res. East Maine.
2.—**Cora C.**, b. July 11, 1867.
3.—**Flora B.**, b. Dec. 8, 1868.
4.—**Wm. E.**, b. May 26, 1871; farmer; res. E. Maine, N. Y.

113 George Robert Twining, son of **52** Mahlon, b. M'ch 8, 1838: m. Eliz. Carman, July 4, 1855, b. Sept. 26, 1835, Broome Co., N. Y. Went to

Wis. about 1861; enlisted in Co. C, 6th N. Y. Reg. known as the "Iron Brigade;" participated in battle of Wilderness and other battles; farmer; res. Mt. Sterling, Crawford Co., Wis.

ISSUE:

1.—**Lucy D.**, b. May 20, 1856; m. Charles R. Baker, May 31, 1877; chil.
2.—**Katie E.**, b. Oct. 31, 1858; d. June 9, 1860.
3.—**Robert L.**, b. May 30, 1861; m. Eliz. Stevenson, July 30, 1881; blacksmith; res. Mt. Sterling, Crawford Co., Wis.; one dau., b. Nov. 20, 1889.
4.—**Emma E.**, b. Aug. 29, 1863; d. Dec. 15, 1874.
5.—**Fred S.**, b. April 26, 1867; farmer.
6.—**Lydia M.**, b. Sept. 6, 1868; d. Oct. 4, 1868.
7.—**Mahlon W.**, b. June 26, 1870; m. Feb. 1888, Abby Girdler; res. Mt. Sterling, Crawford Co., Wis.; one child, Maud, b. Aug. 13, 1889.
8.—**Riley R.**, b. April 28, 1881; farmer.

114 Mahlon J. Twining, son of **52** Mahlon, b. Oct 8, 1841; m. Fanny Galord, of Glenarby, N. Y. Served in the Rebellion one year, N. Y. cavalry. Left Broome Co. about 1867; lived in Ohio two yrs., Wis. four yrs., Iowa a short time, and settled in Tulare Co., Cal., about 1874, where he is engaged in farming and bee keeping; owns a soldier's homestead of 160 acres; res. Hanford, Cal.

ISSUE:

1.—**H. L.**, b. June 4, 1863.

115 Henry L. Twining, son of **52** Mahlon, b. July 17, 1843; m. M. C. McCullick, b. in Wells Co., Ind. Served in the late war; farmer; M. E. Ch.; res. Brookville, Saline Co., Kansas.

ISSUE:

1.—**Dora E.**, b. Feb. 7, 1868.
2.—**John P.**, b. July 2, 1870.
3.—**Charles E.**, b. April 22, 1880.

116 **Jessie Twining**, son of **53** Benjamin, b. April 13, 1834; m. Mary Goodwin, b. Feb. 4, 1831; farmer and dairyman; res. Clark's Summit, Penn.

ISSUE:

1.—**Emma**, b. Nov. 8, 1862; m. Charles Franklin; farmer; res. Clark's Summit.
2.—**Nellie**, b. Oct. 7, 1864; m. Holley Fish; foreman for Western Union Tel. Co.; res. Scranton, Pa.
3.—**Abram**, b. May 3, 1867.
4.—**Maud A.**, b. May 16, 1869.
5.—**Kattie E.**, b. April 23, 1873.

117 **Eli Twining**, son of **53** Benjamin, b. Feb 20, 1836; m. Hannah Taylor, b. Aug. 29, 1836; flagman railroad crossing; res. Scranton, Pa.

ISSUE:

1.—**Hiram E.**, b. May 31, 1862.
2.—**Huldah M.**, b. May 18, 1864.
3.—**Eliz.**, b. April 28, 1866.
4.—**Lydia**, b. March 10, 1868.
5.—**Gertrude**, b. June 17, 1874.

118 **John Twining**, son of **53** Benjamin, b. April 27, 1845; m. Nellie Switzer, b. 1856, d. June 17, 1888; carpenter; res. Scranton.

ISSUE:

1.—**Elmer**, b. 1883.

119 **William Twining**, son of **53** Benjamin, b. M'ch 26, 1847; m. Annie D. Gifford, b. 1852; railroad employe; res Scranton, Pa.

ISSUE:

1.—**Harry B.**, b. 1877.
2.—**Daisy B.**, b. 1879.
3.—**Lula M.**, b. 1881.
4.—**Frank H.**, b. 1886.

120 Horace G. Twining, son of **53** Benjamin, b. July 25, 1854; m. Minnie Sisco, b. 1861; railroad fireman; res. Scranton, Pa.

ISSUE:

1.—**Henry**, b. 1883.

121 Ralph Twining, son of **53** Benjamin, b. Sept. 19, 1858; m. Annie Harris, b. 1861; railroad employe: res. Scranton, Pa.

ISSUE:

1.—**Josanna**, b. May, 1886.

122 Samuel Twining, son of **54** Jacob, b. 1843; m Margaret Rush and has chil.; carpenter; res. Washington, Warren Co., N. J.

123 Malichi Twining, son of **55** Jacob, b. Aug. 3, 1794, d. Sept. 7, 1881, Hancock Co., O.; m. *first*, Oct. 11, 1815, Ann, dau. of **16** Stephen Twining, b. Nov. 28, 1795; d. Sept. 3, 1859, Wrightstown. "A good woman whom he treated badly, so she left him;" m. *secondly*, about 1839, in Hancock Co., Cathrine ———, b. 1812, d. June 7, 1839. He was a farmer and moved to Hancock Co., O., about 1830. The children by his first wife remained in Bucks Co. with their mother.

ISSUE:

1.—**Wilkinson**, b. Sept. 21, 1816, d. July 14, 1849, Bucks; m. Eliz. Jenks, May 23, 1840. *Issue:*
    1. James J., m. and has chil.: *Elmer, Lewis, Mable*, and an infant; farmer; res. Bybery, 23d ward, Phila.
    2. William H., m.; farmer; res. Bybery, 23d ward, Philadelphia, Pa.; dau, Annie.
    3. Anna Mary, m. Jno. Simons; farmer; res. Bybery; chil.: Ella, Harry and Anna.
    4. Charles; res. Philadelphia; unm.
2.—**Mary A.**, b. Sept. 2, 1818; d. inf.
3.—**Elias Stokes**, b. M'ch 5, 1820; m. Eleanor Decoursey, M'ch 15, 1849; farmer; Quaker; res. Penn's Park, Bucks Co., Pa. *Issue:*
    1. Isabella D., b. M'ch 9, 1850, Wrightstown.
    2. Ann M., b. Nov. 10, 1851; d. Jan 5, 1852.
    3. Jane E., b. Dec. 25, 1852; m. Edwin Worthington, Feb. 17, 1876; chil.: *Warren, Morris* and *Eleanor M.;* res. W.
    4. Watson, b. July 13, 1855; m. Amelia Brooks, Nov. 28, 1878, who d. Dec. 1, 1885; res. W.; one child, *Sallie B.*
    5. Albert T., b. Sept. 30, 1857; d. Nov. 19, 1883; m. Emma Worthington, Oct. 1880; res. W.; one child, *Carrie.*
    6. Ellie A., b. June 5, 1860; m. Jno. B. Molloy, Dec. 27, 1879; res. Buckingham, Pa.; chil.: *Bertha* and *Edwin.*
4.—**Maria A.**, b. M'ch 5, 1822; res. Penn's Park; single.
5.—**Watson**, b. Sept. 20, 1825; m. Susan Morgan, Nov. 1849; real estate business, Philadelphia, Pa. *Issue:*
    1. Henry, who m. Jennie S. Hampton, 1874. In business with his father: *Issue:*
        1. Clarence, who d. April 12, 1876.
        2. Russell.
        3. Henry Heman.

ISSUE BY SECOND WIFE:

6.—**John Wm.**, b. May 3, 1840, Hancock Co., O.; res. Leipsic, Putnam Co., Ohio; probably is m. and has issue.
7.—**Caroline**, b. Oct. 28, 1842, Ohio.
8.—**Emily**, b. M'ch 20, 1845.

124 Joseph Twining, son of **55** Jacob, b. Aug. 10, 1800, d. 1859; m. Mary Liverzey, dau. of John and Mary, M'ch 18, 1820, d. April 17, 1877.

Settled in Hancock Co. about 1830. A farmer, justice of peace; postmaster twenty-eight years; teacher; United Brethren.

ISSUE:

1.—**Thomas**, b. Aug. 7, 1823, Bucks Co.; m. Eliz. Bosler, 1847, b. 1828, Cumberland Co., Pa. Moved from Hancock Co., O., to Shelby, Michigan, July, 1867. Served four years in the Rebellion, 21st Ohio, Co. A; farmer. *Issue:*

    1. William, b. 1849; m. Evie Ebert, 1874. Two chil., b. 1874 and 1878 respectively; res. Shelby.
    2. Wesley, b. April 18, 1850; m. Hattie Graves, Jan. 16, 1878; merchant, Shelby; four chil.
    3. Elma, b. Nov. 4, 1851; m. Jary Williams, 1879; merchant; res. Manton, Mich.; two chil.
    4. John, b. June 9, 1855; m. Amelia Burr, of Hart, Mich., Sept. 6, 1881; farmer.
    5. Willibey, b. May 5, 1857; lumberman, Shelby.
    6. Cyrus, b. June 25, 1867.
    7. Millie, b. Sept. 17, 1872.

2.—**Margaret**, b. Jan. 19, 1825.

3.—**Eleazer**, b. 1827; m. *first*, Margaret McBride, 1851, of Canada. Is trying to live with his third wife. A good man; had seven or eight chil.; all d. but three; farmer; res. Findlay, Ohio. *Issue:*

    1. Henry, b. about 1855; lives in Allegan Co., Mich.
    2. Robert, b. about 1866; res. Hancock Co., O.
    3. Jacob, res. Hancock Co., O.

4.—**Mary Ann**, b. July 6, 1830; m. John Farthing; res. Vanliew, Hancock Co., O.; ten children.

5.—**Joseph**, b. Oct. 10, 1834; m. July 3, 1856, Isabella A. Halliwell, b. April 2, 1836; undertaker, Findlay, Ohio, till the time of enlisting in the 21st Ohio Reg., Co. A.; wounded Jan. 1, 1863, in the battle of Stone River and d. Jan. 25 at Nashville, Tenn., hospital No. 12; buried at Nashville one month, and then removed to the Van Horn cemetery, Hancock Co., O.; M. E. Church; wid. m. June 22, 1882, H. W. Davis; res. Fostoria, O. *Issue:*

    1. Amanda, b. May 7, 1857; d. inf.
    2. John H., b. May 25, 1858; telegraph operator; res. Fostoria.

THE TWINING FAMILY. 147

    3. Frank L., b. Dec. 14, 1860; m. Sept. 14, 1882, Minnie Moses, of Kansas, Ohio; telegraph operator; res. Arcada, Ohio. One child, *Cloe*, b. April 5, 1885.

6.—**John**, b. May 1, 1836; d. inf.

7.—**Sarah E.**, b. July 8, 1841; m. Daniel Ramsey; farmer, Dec. 22, 1859; res. Hoytville, Wood Co., Ohio. *Issue:*

    1. William, b. July 31, 1860.
    2. Phebe, b. Aug. 14, 1863.
    3. Albert, b. Nov. 25, 1864.
    4. Sherman, b. June 31, 1867.
    5. John, b. Feb. 18, 1870.
    6. Mary, b. Sept. 6, 1872.
    7. Milton, b. April 4, 1875.
    8. Ella, b. Dec. 6, 1877; d. April 2, 1878.
    9. Charles, b. Jan. 29, 1880.
    10. Cory, b. May 12, 1883.

8.—**Phebe**, m. —— Norrigan; res. Arcada, Ohio.

125 John Twining, son of **55** Jacob, b. M'ch 20, 1802, d. April 25, 1881, W.; m. Mary Lambert, M'ch 15, 1832, b. Aug. 11, 1815; farmer; wid. res. on her farm, a part of the original Twining homestead in Wrightstown.

ISSUE:

1.—**Williamina**, b. July 2, 1833; m. Geo. L. Mahan, Nov. 10, 1853; carpenter; res. W. *Issue:*

    1. Mary E., b. Aug. 17, 1854; m. Franklin Hillborn, April 7, 1878; carpenter; Newtown, Pa.; chil.: four sons.
    2. Phebe Ann, b. Jan. 31, 1856; m. John M. Lee, carpenter; res. Newtown, Pa.; chil.: three sons.
    3. Howard H., b. Dec. 8, 1857.
    4. Adella, b. Sept. 2, 1859.
    5. Lizzie H., b Sept. 28, 1862.
    6. Minnie F., b. Aug. 1, 1864.
    7. Sallie F., b. Dec. 19, 1869.

2.—**N. Lambert**, b. Aug. 4, 1834; m. *first*, Lizzie Roberts, about 1860, who d. 1868; m. *secondly*, Ellie Reeder, 1870, d. 1875-6; m. *thirdly*, Letitia Mathews, 1878; M. E. Church; blacksmith; res. Penn's Park, Pa. *Issue* by first wife:

148   THE TWINING FAMILY.

       1. Elmer E., b. 1862; m. Mary Harvey about 1883; farm laborer; res. Penn's Park, Pa. By second wife:
       2. Graff, b. March, 1872.

3.—**Emily**, b. Oct. 27, 1836; m. Joseph Warner, April 24, 1856; farmer; res. Rush Valley, Bucks Co.; one child, Anna H., b. June 26, 1863.

4.—**Jacob**, b. July 2, 1838, d. Feb. 19, 1864.

5.—**Caroline**, b. July 2, 1842; m. Joseph E. Smith, Nov. 14, 1861, b. Aug. 22, 1828; carpenter and farmer; res. Pineville, Pa. *Issue:*

       1. Warren, b. M'ch 18, 1863; miller; res. Holland, Pa.
       2. John T., b. Jan. 10, 1865; res. Pineville.
       3. Fannie T., b. Jan. 24, 1867; m. Ed. S. Atkinson; farmer.
       4. Lewis C., b. Oct. 13, 1873.
       5. William T., b. Nov. 15, 1875.

6.—**William**, b. Aug. 17, 1844; m. Lettie Firman, Oct. 2, 1873; res. Philadelphia, Pa. *Issue:*

       1. Harry A., b. Sept. 13, 1874.
       2. Mary E., b. July 25, 1878.

7.—**Charles**, b. Oct. 22, 1847; m. Sallie Blaker, Sept. 18, 1877; farmer; res. W. *Issue:*

       One son, J. Augustus, b. June 14, 1872.

8.—**Hannah F.**, b. Feb. 22, 1851; m. John Kennedy, Oct. 31, '72; blacksmith; res. Buckmanville, Pa. *Issue:*

       1. Mary, b. June 29, 1873; d. June 9, 1875.
       2. Annie, b. May 19, 1876.
       3. Laura, b. M'ch 17, 1879.
       4. Ella B., b. Nov. 13, 1883.
       5. Charles T., b. Aug. 7, 1885.

126 James Twining, son of **55** Jacob, b. Oct. 10, 1808, W.; deceased; m. Elizabeth Staley, 1829. Came to Hancock Co., Ohio, with his brothers and sister about 1830; farmer; belonged to the U. B. Church and could "shout as loud as any man in the county;" wid. resides near Findlay, O.

ISSUE:

1.—**Mary A.**, b. Dec. 13, 1831; res. Hancock Co.
2.—**Angeline**, b. Dec. 29, 1832; m.; res. Mich.

3.—**Allen**, b. M'ch 11, 1834; m. Rachel Amond; res. Burnip's Corners, Allegan Co., Mich. Probably has children.
4.—**Charles**, d. inf.
5.—**Isabella**, b. Oct. 20, 1837; res. Hancock Co.
6.—**Emeline**, b. July 11, 1841; res. Hancock Co.
7.—**Elizabeth**, b. Oct. 2, 1842.
8.—**Lydia**, b. May 5, 1844.
9.—**Sarah**, b. Oct. 18, 1845.
10.—**Almira**, b. Sept. 20, 1848.
11.—**William J.**, b. June 21, 1850; m. Kate Snyder; res. Findlay, Hancock Co.
12.—**Harriet**, b. Sept. 3, 1853.
13.—**Anna**, b. July 11, 1856.

127 Jacob Twining, son of **55** Jacob, b. April 12, 1812; m. Elizabeth Adams (wid.) about 1836. She d. Sept. 5, 1874, aged 65 years, 10 months; buried in family burial grounds, Salem Cemetery, Hancock Co., O., where three of her chil. are buried, the others in the Van Horn cemetery; res. Carey, Wyandot Co., formerly of Hancock Co., the only surviving member of the old stock who came to Ohio.

ISSUE:

1.—**Phebe**, b. about 1837; m. *first*, Filtenberger, from whom she parted; m. *secondly*, and went West, where she died.
2.—**Frances B.**, b. Dec. 1839, d. Jan. 7, 1874; m. David Frasier; left one dau., who is married.
3.—**Henry Clay**, b. about 1842, d. in Mich., 1888; buried in Van Horn Cem.; m. Miss Ranchler, now in insane asylum; six children, three of whom are dead.
4.—**Harriet**, d. when young.
5.—**John A.**, b. May, 1849; d. June 2, 1872.

128 Ralph L. Twining, son of **55** Jacob, b. July 23, 1820, W., d. Oct. 1, 1870, of tapeworm, W.; m. M'ch 2, 1843, Annie, only dau. of Samuel and Mary

Heaston, of Washington's Crossing, Pa., b. June 28, 1813, d. Jan. 21, 1885. "After the completion of his studies he became a wheelwright, and many fine carriages and sleighs still testify to his superior workmanship. He inherited the old homestead at W., which remains in the possession of his grand daughter. A noble man, kind and genial. His wife was a Christian woman of great energy, and at his death successfully managed the farm."

ISSUE:

1.—**Samuel H.**, b. M'ch 16, 1844; m. Feb. 16, 1862, R. Jennie Homer. Soon after he entered the army, Co. "A," 186th Reg. Pa. Infantry; d. of measles in the field hospital, M'ch 20, 1864. *Issue:*
    1. Maud C, b. Nov. 24, 1862; m. Samuel K., eldest son of Samuel and Eliz. Wismer, a German family, M'ch 4, 1885. She was left an orphan while yet an inf. and adopted by her grand mother. One child, Samuel K., b. Aug. 18, 1887.
    2. Elmer, b. 1864; res. Philadelphia; single.

2.—**John H.**, b. Feb. 21, 1851; d. 1852.

3.—**Ralph L.**, b. Nov. 6, 1854; d. 1874; unm.

129 Silas Twining, son of **56** John, b April 26, 1802, Bucks Co.; d. Jan. 12, 1854; res. a short distance south of Findlay, Ohio; m. Letitia Harrold.

ISSUE:

1.—**John**, b. Feb. 6, 1829; d. Dec. 29, 1848.

2.—**Charles**, b. Feb. 10, 1833; d. in the army at Knoxville, Tenn., Sept. 14, 1864, from sunstroke.

3.—**Amos**, b. July 28, 1837; m. Nov. 18, 1858; served three years in the 18th U. S. A. Infantry, Ohio; gardener and fruit grower; Baptist; res. Haskins, Wood Co., Ohio, where he has res. since 1860. *Issue:*
    1. Lucinda, b. May 5, 1860; m.
    2. William H., b. May 13, 1862; m.
    3. Elam B., b. July 18, 1866; m.

THE TWINING FAMILY.   151

    4. Susan A., b. May 18, 1868.
    5. Harvey A., b. May 13, 1871.
    6. Jennie M., b. Oct. 30, 1873.
    7. George W., b. Aug. 12, 1875.
    8. Wilber R., b. Sept. 3, 1877. These chil. all res. Haskins, Ohio.

4.—**Eli**, b. Feb. 23, 1839; d. in army at Knoxville, Tenn., Aug. 26, 1864, of typhoid fever.

5.—**Francis M.** b. Nov. 21, 1842; m. Theadocia A. Apger, April 8, 1866, b. M'ch 10, 1842, N. J.; gardener; Baptist; res. Haskins, Ohio. *Issue:*
    1. Mary L , b. April 29, 1868.
    2. Francis E., b. Jan. 22, 1872.
    3. Annie E., b. Nov, 22, 1875; d. April 25, 1877.
    4. Jay W., b. M'ch 2, 1877.
    5. Lula, b. Nov. 27, 1883; d. Jan. 2, 1884.

6.—**Alvin**, b. April 25, 1847; d. in army near Washington, D. C., Aug. 24, 1864, of measles.

7.—A son, d. young.

8.—A dau., d. young.

130 Jonathan Twining, son of **57** Joseph, b. Nov. 19, 1809, d. Sept. 29, 1858, Mauch Chunk, Pa.; m. Susan Balliel, Sept. 17, 1836, d. Feb. 21, 1856, b. Aug. 31, 1819. General store at Beaver Meadows, Pa.; thought to be wealthy at his death, but claims brought against estate, never heard of, which were paid; left him in moderate circumstances. Noted for his upright dealings and a most highly respected citizen.

ISSUE:

1.—Infant dau.; d. April 17, 1837.

2.—**Mary Jane**, b. Jan. 17, 1839; m. Elias D. Thompson, M'ch 26, 1857: she d. July 28, 1887, Mauch Chunk, Pa. *Issue:*
    1. Ellen, b. April 24, 1857; d. Jan. 8, 1863.
    2. Willie, b. Nov. 13, 1858; d. Sept. 8, 1859.
    3. Laura, b. Feb. 13, 1859, M. C.; m. Charles A. Crook, June 11, 1884; he d. April 20, 1885; baggagemaster on D. L. & W. R. R.; wid. res. Scranton, Pa.
    4. Edgar, b. Dec. 1, 1861; m. Alma Schock, Oct. 14, 1886; res. Allentown, Pa.

152 THE TWINING FAMILY.

3.—**Edgar**, b. Oct. 10, 1840. Is one of the best known men in Carbon Co., Pa. Has been connected with First National Bank, Mauch Chunk, for twenty years; now cashier; largest real estate owner in E. Mauch Chunk; at one time treasurer of Carbon county; single; Republican.

4.—**George W.**, b. April 6, 1842; m. Sallie Slutter, of Slatington, Pa., April 24, 1865. Supt. of the Lehigh and Susquehanna division of the New Jersey Central R. R.;

LIEUT. HARRY LA VERNE TWINING.
Of Riverside Pub. Schools. Lieut. Co. C, 9th Cal. Regt. (See page 128.)

began with company as supervisor of bridges; has one of the prettiest homes in East Mauch Chunk and has large grounds overlooking the Lehigh river and valley, on which he contemplates building a $20,000 residence; Republican. *Issue:*

    1. Mamie, b. April 28, 1866; d. April 22, 1874.
    2. William, b. July 8, 1867; attending Cornell University, N. Y.
    3. Sallie, b. Sept. 17, 1869.
    4. Georgiana, b. Oct. 13, 1872.

5.—**John**, b. April 16, 1845; d. inf.

THE TWINING FAMILY.

6.—**Eveline**, b. July 7, 1846; d. Nov. 2, 1847.
7.—**Antoinette L.**, b. Dec. 21, 1850; m. Wm. S. Walter, Dec. 4, 1874; clerk, Central R. R. of N. J.; Episcopalian; res. Mauch Chunk.
8.—**Amanda**, b. Nov. 1, 1850; M. E. church; single; res. Scranton, Pa.
9.—**Alfred W.**, b. Feb. 12, 1853; reporter and foreman of *Sunday Free Press*, Scranton; owner of store Vorhis & Twining; owns one of the finest properties in S., and of local note for cultivation of flowers; Democrat; M. E. Ch.
10.—Inft. son, d. July 31, 1854.
11.—**Thompson J. D.**, b. Jan. 4, 1856; d. Feb. 27, 1856.

131 Hallowell S. Twining, son of **58** Watson, b. April 5, 1824, d. Dec. 6, 1885; drowned when in a vessel in the Atlantic ocean; m. Jane Williams; farmer; res. Horsham, Montgomery Co., Pa.

ISSUE:

1.—**Fannie**, b. Jan. 24, 1849; m. Samuel J. Thomson, M'ch 20, 1878, who d. Dec. 13, 1882; one child, Caroline, b. Nov. 29, 1879.
2.—**Harriet**, b. April 7, 1852; d. July 25, 1853.
3.—**William A.** (twin), b. April 7, 1856; m. Laura Knight, M'ch 12, 1879; one child, Walter, b. May 31, 1880; res. Philadelphia, Pa.
4.—**Watson** (twin), b. April 7, 1856; d. Oct. 8, 1859.
5.—**Laura**, b. Dec. 9, 1860; m. John R. Tyson, Oct. 7, 1885; one child, Warren, b. Sept. 6, 1886.
6.—**Watson**, b. Nov. 3, 1863; res. Philadelphia.
7.—**Russell**, b. Nov. 22, 1865; lives on his father's farm at Horsham, Penn.
8.—**Silas H.**, b. Dec. 9, 1871; res. Philadelphia.

132 Elias B. Twining, son of **58** Watson, b. Sept. 26, 1832; m. Charlotte Tyson; he d. July 8, 1862; buried in Warminster Tp.; one son, Edroy, b. Nov. 1860.

133 Silas Twining, son of **59** Silas, b. M'ch 21, 1848; m. Annie T., dau. of James and Jane Vanartsdalen, June 17, 1875; clerk; res. Philadelphia.

ISSUE:

1.—**Eugene**, b. M'ch 17, 1876.
2.—**Howard V.**, b. M'ch 10, 1885.

134 Uriah R. Twining, son of **60** William, b. Oct. 16, 1831, on the "old homestead, Warwick Tp.;" m. Juliann Vanartsdalen, Nov. 30, 1854, b. Feb. 16, 1836; farmer; res. Trevose, Bucks Co., Pa.

ISSUE:

1.—**Elwood**, b. Aug. 19, 1855; m. Letetia Ridge, Feb. 1, 1877; res. Bensalem, Bucks Co., Pa.
   1. Harry, b. Aug. 13, 1877.
   2. Emma, b. Sept. 24, 1880.
   3. Maggie, b. M'ch 22, 1887.
2.—**William Thos.**, b. April 19, 1857; m. Maggy Lindsey, June 22, 1882; res. Moreland, Montg. Co., Pa. *Issue:*
   1. Beulah E., b. Jan. 17, 1884.
3.—**Uriah**, b. Jan. 19, 1859; res. Bensalem; single.
4.—**Beulah**, b. Feb. 1, 1861; m. John Page, Dec. 24, 1885; res. Southampton, B. Co., Pa.
5.—**Silas**, b. Oct. 22, 1864; res. Northampton, B. Co., Pa.; single.

135 William W. Twining, son of **60** William, b. Jan. 17, 1844, d. July 30, 1880, Trenton, N. J.; insane asylum, caused by spiritualism; m. Mary A. Van Horn, M'ch 3, 1870; res. Wrightstown, Pa.

ISSUE:

1.—**Emma L.**, b. May 28, 1871.
2.—**Lydia V.**, b. Nov. 26, 1872.
3.—**Rebecca E.**, b. Sept. 2, 1874.
4.—**Walter C.**, b. Feb. 6, 1880.

136 D. Hallowell Twining, son of **61** Isaac, b. Aug. 29, 1828; m. Alice P. Baynes, of Baltimore, Md., Dec. 14, 1865, b. 1836 and d. M'ch 1, 1876; farmer; res. Greenwood, Baltimore Co., Md.

ISSUE:

1.—**Joseph B.**, b. M'ch 4, 1867; res. Upper Cross Roads, Harford Co., Md.
2.—**Horace B.**, b. Sept. 13, 1868; d. April 11, 1871.
3.—**Isaac**, b. May 24, 1871.
4.—**B. Franklin**, b. Oct. 5, 1874; d. Dec. 2, 1874.

137 Horace B. Twining, son of **61** Isaac, b. Sept. 15, 1832; m. Fannie Ashton, Feb. 20, 1872, b. April 15, 1843. From 1861 to 1869 on the Pacific slope; farmer; birthright Quaker; res. Forest Hill, Harford Co., Md.

ISSUE:

1.—**Mary Ann**, b. M'ch 15, 1873.
2.—**Albert B.**, b. Jan. 10, 1878.

138 B. Franklin Twining, son of **61** Isaac, b. Oct. 12, 1837; m. Mary C. Nippes, of Phila., Pa., July 21, 1864; he d. Jan. 3, 1880, Phila. "Left Friends and joined Lutherans."

ISSUE:

1.—**R. Barclay**, b. June 16, 1867; res. Philadelphia.
2.—**Mary Ella**, b. Dec. 29, 1873; d. July 17, 1874.

139 Charles H. Twining, son of **62** Thomas, b. Nov. 12, 1853; m. Jan. 9, 1873, Mary A. Savidge, b. in McLean Co., Ill., Jan. 26, 1854. 'Born and lived in the original house his father built upon a farm of

280 acres, to Oct. 1884, when he removed to Little River, Kansas, where he is engaged in farming, stock raising and real estate.

ISSUE:

1.—**Musetta**, b. Dec. 15, 1875.
2.—**T. J.**, b. Sept. 6, 1877.
3.—**Earl**, b. Jan. 11, 1879; d. Jan. 1, 1885.
4.—**Ted R.**, b. Jan. 17, 1883.

140 John Twining, son of **63** Jacob, b. Oct. 10, 1839; m. Kath. S. Frankinfield, Jan. 3, 1860; res. Folsom, Custer Co., Dakota, where he moved 1886.

ISSUE:

1.—**Rachel**, b. M'ch 30, 1862; d. Dec. 11, 1886.
2.—**Samuel**, b. Jan. 17, 1865.
3.—**Charles**, b. M'ch 20, 1866.
4.—**Annie**, b. M'ch 16, 1869.
5.—**Susan**, b. April 16, 1871.
6.—**John D.**, b. Nov. 3, 1874.
7.—**Jane E.**, b. M'ch 26, 1877.
8.—**George W.**, b. May 5, 1879.

141 Isaac H. Twining, son of **63** Jacob, b. Feb. 23, 1845; m. Mary A. Whit, Dec. 25, 1868, b. Highland, Wis., M'ch 24, 1846. Moved to Neb. 1879; blacksmith and wagon maker; res. Bazile Mills, Neb.

ISSUE:

1.—**Isaac W.**, b. Nov. 5, 1869; d. M'ch 6, 1886.
2.—**Sarah M.**, b. Dec. 4, 1871.
3.—**William**, b. April 5th, 1874; d. inf.
4.—**Arthur M.**, b. April 1, 1875.
5.—**Mont. H.**, b. Dec. 5, 1877.
6.—**Elsie F.**, b. Jan. 24, 1880.
7.—**Ellen**, b. July 31, 1882.
8.—**Earmer M.**, b. Sept. 4, 1884.

142 John W. Twining, son of **64** Abbott C., b. June 11, 1837; m. Mary Ellen Briggs, Feb. 1, 1868; Friends; farmer; res. Rush Valley, Bucks Co., Pa.

ISSUE:

1.—**Edward W.**, b. Oct. 18, 1873; d. M'ch 1, 1874.
2.—**D. Walker**, b. Nov. 26, 1877.

143 Thos. Chalkley Twining, son of **64** Abbott C., b. Feb. 12, 1844; m. Mary E. Kirk, Jan. 24, 1867; pump manufacturer; res. Mozart, Bucks Co., Pa.

ISSUE:

One child, Joseph W., b. M'ch 7, 1868; d. inf.

144 David R. Twining, son of **65** Isaac, b. Nov. 29, 1851; m. Hannah Kyle, Nov. 1870, d Aug. 23, 1883, aged 32 years; blacksmith; res. Southampton, Bucks Co., Pa.

ISSUE:

1.—**Sallie E.**, b. May 20, 1872.
2.—**Susan F.**, b. Aug. 23, 1874.
3.—**Lizzie**, b. Aug. 13, 1876.
4.—**David R.**, b. Dec. 23, 1878.
5.—**John**, b. M'ch 28, 1881.

145 Howard L. Twining, son of **67** Henry M., b. July 13, 1850; m. Mary E. Cooper, Jan. 8, 1873; train hand; res. Philadelphia.

ISSUE:

1.—**Anna L.**, b. Oct. 24, 1873.
2.—**Harvey**, b. M'ch 21, 1875.
3.—**May**, b. Dec. 1, 1876; d. Sept. 30, 1882.
4.—**Walter**, b. July 3, 1879.
5.—**Aimee B.**, b. Feb. 1, 1884.
6.—**Achsah V.**, b. Dec. 3, 1886.

146 Jonathan R. Twining, son of **68** Cyrus B., b Sept. 10, 1852; m. Bell Warner, Jan. 15, 1879; farmer; res. Pineville, Bucks Co., Pa. Also in pork business, Philadelphia.

ISSUE:

One child, F. Cyrus, b. June 15, 1879.

147 Wilmer A. Twining, son of **68** Cyrus B., b. April 17, 1865; m. Lottie Vandegrift, Oct. 1886; farmer; res. Pineville, Pa.

ISSUE:

One child, Franklin M., b. Sept. 15, 1888.

148 William H. Twining, son of **69** Amos H., b. Feb. 8, 1845; m. Mary E. Echart, Jan. 1, 1873; res. Richboro, Bucks Co., Pa.

ISSUE:

1.—Clarence B., b. May 15, 1874.
2.—Ethel M., b. M'ch 27, 1876.
3.—Jennie B., b. May 15, 1887.

149 John Twining, son of **69** Amos H., b. Dec. 14, 1849; m. Mary E. Slack, M'ch 8, 1879; farmer; res. Richboro, Bucks Co., Pa.

ISSUE:

1.—Emma H., b. July 8, 1880.
2.—Amos H., b. May 7, 1887.

150 Edwin Twining, son of **71** Croasdale, b. Dec. 1837; m. Hannah A. Iredell, Nov. 1864; in U. S. Cavalry during the Rebellion; commission merchant; res. Chehalis, Washington; family live at Davis Grove, Pa.

THE TWINING FAMILY.

ISSUE:

1.—**J. Howard**, b. Dec. 4, 1865.
2.—**William P.**, b. June 11, 1868.
3.—**Anna Mary** (" Dollie "), b. Feb. 13, 1871.
4.—**Ida**, b. July 8, 1873.
5.—**Ellie B.**, b. Sept. 23, 1875.
6.—**A. Iredell**, b. Sept. 2, 1878.

151 Kinsley Twining, son of **73** Alexander C., b. 1832, New Haven; grad Yale, 1853; Congregational minister, Doctor of Divinity; on editorial staff of the New York *Independent*; m. *first*, 1861, Mary K. Plunkett, of Hinsdale, Mass., d. 186–; m. *secondly*, 1870, Mary Gridley, dau. of A. D. Gridley, of Clinton, N. Y.; res. Morristown, N. J.

ISSUE BY SECOND WIFE:

1.—**Edith**, b. 1872.
2.—**Alin**, b. 1877.
3.—**Kinsley**, b. 1879.

152 Edward H. Twining, son of **74** William, b. Lowell, Mass, Oct. 3, 1833; grad. Wabash Coll., 1852; private 33th Ill. In'f'y, 1861; Capt. 1862; Capt. A. D. C. 1864; Prof. Chem. Washington and Jefferson Coll., 1866; University Minn., 1869: Prof. Latin, Univ. Mo., 1872; resigned 1877. Afterward connected with the St. Louis high schools; since 1882 the Mississippi River Commission, New York city; Episcopal church; m. Aug. 6, 1860, Harriett, dau. of C. S. and Catharine (Leavenworth) Sperry, of Waterbury, Ct.; she d. Dec. 15, 1876, Columbia, Mo. "A brilliant man of a very retentive mind."

ISSUE (Four chil. d. inf.):

1.—**Jane L.**, b. Feb. 6, 1866; stenographer; small and slight in stature; educated in St. Louis (Mo.) city schools; fond of business; single; res. Waterbury, Ct.
2.—**Almira C.**, b. Nov. 13, 1867, Washington, Pa.; lively and brilliant in conversation, lithe and strong and "has the 'set' look of the Twinings;" assistant in the Bronson Ribbie Library, Waterbury, Ct.; single.
3.—**Wm. Edward**, b. Nov. 19, 1874, Mo.; small and slender; accomplished as a skater, billiardist and bicycleist; res. Waterbury, Ct.

153 Charles O. Twining, son of **74** William, b. Crawsordsville, Ind., Sept. 28, 1845; m. Dec. 23, 1873 Ann Campbell, who d. April 6, 1875; Sec. French Window Glass Manufacturing Co., St. Louis, Mo.; Cong. Ch.

ISSUE:

1.—**Ralph C.**, b. St. Louis, Feb. 28, 1875.

154 George A. Twining, son of **75** Alfred C., b. M'ch 15, 1841, Lansingsburgh, N. Y.; m. Jennie Byers, b. in Eng. and d Jan. 1, 1882. Served in in 7th Conn., Co. A, in the late war. Came to Chicago after the great fire; dealer in coal; familiarly known as "Old Rockey;" res. Chicago, Ill.

ISSUE:

1.—**Lillie**, b. Aug. 12, 1872.
2.—**George**, b. June 10, 1874.
3.—**Mollie**, b. M'ch 10, 1876.

155 William F. Twining, son of **76** Alexander H., b. Nov. 18, 1851; m. Eva M. Carpenter, of Strongsville, Ohio, Dec. 23, 1874; machinist; res. 151 Abram street, Cleveland, Ohio.

ISSUE:

1.—**Charles F.**, b. Nov. 18, 1875, Strongsville.
2.—**William A.**, b. M'ch 16, 1877, Cleveland.
3.—**Laura M.**, b. April 9, 1880, Cleveland.
4.—**Roscoe D.**, b. Dec. 23, 1883, S.
5.—**Blanche M.**, b. Aug. 3, 1886, C.

156 John Twining, son of **77** William, b. Jan. 6, 1816, T., d. Sept. 6, 1876, Copenhagen, N. Y., where he res.; m. Eveline R. Smith, 1844, b. Oct. 14, 1821, Wilna, N. Y.; farmer and Free Thinker.

ISSUE:

1.—**Jno. S.**, b. 1845; d. inf.
2.—**John S.**, b. Jan. 27, 1847; m. Sept. 21, 1869, Mary E. Patten, b. M'ch 10, 1848, Ava, N. Y.; farmer and lumber dealer; Free Thinker; res. Copenhagen, N. Y. He has in his possession a powder horn made from ox horn carried by one of the first Cape Cod family; the initials on bottom are " W. T. Jr." *Issue:*
   1. Carrie E., b. Feb 8, 1872.
   2. Cora M , b. Feb. 22, 1880.
3.—**Ovanda M.**, b. June 22, 1848; m. Warren Hawn, Starkville, N. Y.; two chil.
4.—**William J.**, b. June 26, 1860; farmer; Free Thinker and single; res. Copenhagen.

157 William F. Twining, son of **77** William, b. Aug. 17, 1820, N. Y., d. July 20, 1886: m. *first*, Martha M. Taylor, of Wilna, N. Y., Feb. 19, 1846, b. Saratoga Co., N. Y., 1824, d. July 29, 1879, Morrison, Ill.; m. *secondly*, Miss Nellie Rook, of M., M'ch 26, 1882. Came to Morrison, Ill., from Jefferson Co., N. Y., 1863; farmer; Universalist; filled important township offices; wid. res. Unionville, near Morrison.

ISSUE BY FIRST WIFE:

1.—**Mary Frances**, b. Jan. 2, 1846; d. May 12, 1849.
2.—**William E.**, b. June 8, 1848; m. M'ch 23, 1887, m; Ida M. Baker, of M.; farmer; res. Lake City, Iowa.
3.—**Mary F.**, b. Oct. 10, 1849; m. Orrin M. Bent, Jan. 16, 1866, formerly of Denmark, N. Y.; a wealthy farmer; res. Morrison, Ill. *Issue:*

    One dau., Lottie F., b. Dec. 6, 1866; m. Richard Tilton, of Whiteside Co., Ill.

4.—**Fred A.**, b. April, 1857; d. M'ch 5, 1862.
5.—**Florence A.**, b. June 11, 1860; m. Absalom C. Venumn, July 1, 1880, of Exeter, Neb.; engaged in banking business, Stratton, Neb. *Issue:*

    1. Ethel, b. June 12, 1881.
    2. Stella, b. July 19, 1883.

ISSUE BY SECOND WIFE:

6.—**Grace B.**, b. Nov. 18, 1884.

158 Alfred W. Twining, son of **77** William, b. Sept. 3, 1822; m. *first*, Jennette Fargo, Feb. 16, 1840; m. *secondly*, Miranda Gibbs, Sept. 12, 1859; farmer; res. So. Champion, N. Y.

ISSUE BY FIRST WIFE:

1.—**Nelson P.**, b. Oct. 13, 1848; m. Ada Lewis, Feb. 6, 1869, b. July 10, 1849; res. So. C.; farmer. *Issue:*

    1. Lewis A., b. Oct. 10, 1870. Will probably enter West Point U. S. M. A., June, 1890.
    2. Belle E., b. July 6, 1872.

2.—**Emogene C.**, b. 1851; m. James O. Waldo, Westernville, N. Y.; one dau.

ISSUE BY SECOND WIFE:

3.—**George**, b. Dec. 24, 1860; m. Lottie Cramer, Feb. 13, 1884. One child, Gleepe A., b. Oct. 23, 1885; res. Copenhagen.
4.—**Ada E.**, b. Sept. 1863; m. Henry E. Chickering, June 18, 1884; res. Copenhagen, N. Y.

159 Elphonzo Twining, son of **78** Elijah, b. June 8, 1818, T.; m. *first*, Eliza A. Cone, M'ch 5, 1850, b. July 14, 1823, d. Sept. 14, 1865; m. *secondly*, Sarah C. Wheldin, of Williamston, Mass., Feb. 19, 1867, d. Oct. 10, 1867, m. *thirdly*, Annie E. Gates, of Hoboken, N. J., April 29, 1869, d. April 8, 1885, of Bright's disease; farmer; moved to Sandisfield, Mass., his present res., 1850.

ISSUE BY FIRST WIFE:

1.—**Belle E.**, b. Dec. 27, 1850; m. Charles C. Cooper, Oct. 10, 1871; three chil.; res. Chester, Mass.
2.—**Orlow C.**, b. M'ch 14, 1853; m. Emma Merrell, Nov. 29, 1876; farmer; res. Sandisfield; one son, *William A.*, b. M'ch 28, 1882.
3.—**Preston M.**, b. Nov. 20, 1855; d. Feb. 24, 1857.
4.—**Flora E.**, b. M'ch 11, 1857; res. Chester, Mass.
5.—**Clinton B.**, b. M'ch 8, 1860.
6.—**Sarah C.**, b. Dec. 14, 1862; d. May 6, 1885.

ISSUE BY SECOND WIFE:

7.—**Winthrop B.**, b. May 21, 1871.
8.—**Charles G.**, b. July 28, 1872.
8.—**Clive I.**, b. Oct. 1, 1874; d. Sept. 11, 1879.
10.—**Geneviere C.**, b. Nov. 11, 1877.

160 Joseph Twining, son of **78** Elijah, b. April 23, 1820, T.; m. Henrietta M., dau. of Rev. Joel Talcott and Lois, dau. of **37** Judah Twining, Nov. 20, 1851, b. Aug. 29, 1830, Wellington, Ohio; farmer; res. Colebrook, Ct.

ISSUE:

1.—**Albert T.**, b. Feb. 20, 1854, Sandisfield; m. Mary J. Coy, April 2, 1878; farmer; res. Norfolk, Ct.

2.—**William J.**, b. June 23, 1855, S.; m. Lucy Morehouse, Dec. 5, 1881, who d. May 22, 1883; res. Colebrook. One child, Bessie H., b. April 28, 1883.
3.—**Katie E.**, b. April 11, 1878, S.; res. Colebrook.
4.—**Burton P.**, b. July 27, 1859, S.; m. Mazzie Ramsey, of New Britain, Ct., April 13, 1882; expressman; one child, Joseph B., b. May 9, 1883.
5.—**Charles J.**, b. Sept. 24, 1861; res. New Britain, Ct.; farmer.
6.—**Eugene R.**, b. Sept. 4, 1863; res. Colebrook; farmer.
7.—**Frank F.**, b. Dec. 10, 1865; d. C., Aug. 27, 1879.
8.—**Frederick H.**, b. Jan. 2. 1867; res. C.; farmer.
9.—**Lois E.**, b. Sept. 12, 1869; d. Sandisfield, Oct. 18, 1872.

161 Orlandon Twining, son of **78** Elijah, b. Sept. 30, 1821; m. Lucy E. Ervin, April 6, 1853, b. Nov. 6, 1830; farmer; res. Copenhagen, Lewis Co., N Y.

ISSUE:

1.—**Charles E.**, b. Feb. 11, 1854; d. Nov. 22, 1856.
2.—**Bevel E.**, b. May 1, 1858; "traveling agent in the Western States;" single.
3.—**Cassius H.**, b. July 15, 1861; m. Feb. 8, 1888, Minnie E. Elmer, of Lowville, N. Y., b. M'ch 6, 1866; farmer; res. Lowville, Lewis Co., N. Y.
4.—**Clinton J.**, b. Nov. 3, 1867; m. Aug. 15, 1888, Ida A. Spencer, of Denmark, N. Y., b. May 31, 1867; farmer; res. Copenhagen, N. Y.

162 Samuel M. Twining, son of **78** Elijah, b. Feb. 9, 1824, d Aug. 23, 1887; m. Oct. 16, 1850, Harriet, dau. of John Gates. Moved from Tolland, Mass., to East Hartland, Ct , where all but one of his chil. were born; farmer; family Cong.

ISSUE:

1.—**John G.**, b. July 14, 1851; m. Nov. 14, 1878, Etta I. Hoskins (dau. of David); merchant; res. Waterbury, Ct. *Issue:* (See engraving.)

1. Addie, b. April 18, 1880.
   2. Alice, b. Jan. 18, 1882.
   3. Emma, b. Aug. 7, 1884.
2.—**Hattie**, b. Dec. 29, 1853; m. May 31, 1876, Wm. J. Colton.
3.—**Nellie I.**, b. Feb. 22, 1856; d. Sept. 10, 1888; m. Frank Morgan, Dec. 1876; res. East Granby, Ct.

JOHN G. TWINING.

4.—**Austin H.**, b. Sept. 6, 1859; m. Feb. 9, 1885, Mrs. Mary Raidick; hotel keeper; res. Cairyville, N. Y.
5.—**Emma A.**, b. Jan. 12, 1861; m. Frank Johnson, Dec. 25, 1879; res. Hartford, Ct. *Issue:*
   1. Etta I., b. Nov. 6, 1880.
   2. Minnie J., d. Dec. 28, 1888.

163 Lucius Twining, son of **78** Elijah, b. Aug. 8, 1827, T.; m. Mary E. Jackson, June 5, 1850; res. East Granville, Mass.

ISSUE:

1.—**Frank L.** (twin), b. June 5, 1852; m. Jessie Andrews, Dec. 16, 1880. *Issue:*
   1. Sybil E., b. Sept. 12, 1881.

       2. Lewis F., b. July 18, 1883.
       3. Harry K., b. Jan. 14, 1886.

2.—**Fred L.** (twin), b. June 5, 1852; m. *first*, Jennie L. Miller, April, 1876; she d. Feb. 1877; m. *secondly*, Hattie Cowan, 1882.

3.——, b. Feb. 14, 1854; d. Feb. 14, 1868.

4—**Emma A.**, b. Oct. 22, 1858; d. June 29, 1875.

5.—**J. J.**, b. Jan. 4, 1864.

6.—— b. April, 1871; d. inf.

164 Samuel R. Twining, son of 79 Hiram,

b. Jan. 30, 1831; d Nov. 12, 1881, Fiatt, Fulton Co., Ill.; m. Sarah E. Overstreet, Sept. 13, 1854, b. and raised in Mitchellsburgh, Boyle Co., Kentucky. Came to Ill. 1870. Samuel and his brother Henry N. went with the famous "squirrel hunters" to capture the rebel, Jno. Morgan; carpenter and farmer.

          ISSUE:

1.—**Clarence W.**, b. June 6, 1855; m. Hattie G. Quick, M'ch 8, 1883; farmer; res. Fiatt, Ill.

2.—**Edwin H.**, b. Sept. 18, 1859; farmer; res. Fiatt; m. Grace G. Fluke, Jan. 20, 1887.

165 Nelson B. Twining, son of 82 Philander

F., b Nov. 16, 1832; m. M. Eliz., dau. of Alfred and Emoline (Torry) Webb, Tolland, Sept. 10, 1854; farmer; res. New Boston, Mass.

          ISSUE:

1.—**Edson P.**, b. Sept. 25, 1856; m. Sept. 30, 1875, Anna I. Fosdick, b. 1854, d. Oct. 17, 1888. *Issue:*

    1. Bessie A., b. Feb. 23, 1876.
    2. Alice A., b. June 4, 1878.
    3. Lena M., b. Aug. 10, 1881.
    4. Pearl E., b. Sept. 25, 1883.
    5. Lois E., b. Sept. 18, 1885; d. Nov. 11, 1888.
    6. George E., b. Jan. 5, 1888; d. Oct. 29, 1888.

2.—**Howard W.**, b. May 14, 1860; m. Jan. 28, 1890. Hattie G. Kilfoil; res. N. Boston.
3.—**Lois Etta**, b. July 31, 1861; m. Aug. 20, 1886, Dempster Hamlin, of Middletown, Ct.
4.—**Catherine E.**, b. Sept. 26, 1871.

166 Homer P. Twining, son of **82** Philander F., b. Nov. 9, 1839; educated Williston Sem.; held various town offices and represented his dis. in the Mass. Leg. 1882; farmer, at New Boston, Mass.; m. Mary B. White, dau. of Horace and *Susan (Wolcott), Jan. 1, 1861; grad. C. L. S. C. and has many deeds and documents of the early Twining families of Eastham and Tolland; Cong. Ch. Mr. T. is now manager "N. Y. Hotel Journal," N. Y. city.

ISSUE:

1.—**Clifford H.**, b. Nov. 22, 1866; d. Jan. 27, 1868.
2.—**Clifford H.**, b. April 22, 1869.

* Niece of Edward Wolcott, who m. Susannah, dau. of **18** Elijah Twining and of Edward, who m. Anna of **17** Thomas Twining.

167 Lewis S. Twining, son of **83** Merrick, b April 6, 1833, Berlin, O.; m. Jan. 26, 1854, Philena C. Moore, b, Dec. 1, 1835; she d. May 12, 1885; res. Granville, O.

ISSUE:

1.—**Carper W.**, b. Jan. 10, 1855; d. inf., G.
2.—**Leota O.**, b. Oct. 13, 1856, Otho, Iowa.
3 —**Arthur M.**, b. Sept. 12, 1859; d. Jan. 30, 1864, Iowa.
4.—**Lucine O.**, b. July 11, 1862; d. April 25, 1865, Iowa.
5.—**Minnie A.**, b. Aug. 7, 1865; d. inf.
6.—**Nettie E.**, b. Dec. 18, 1873, Otho, Iowa.

168 Edward W. Twining, Jr., son of **83** Merrick, b. May 8, 1836; d. Jan. 31, 1888, of consumption, Fort Dodge, Iowa; m. Oct. 2, 1861, *Matilda Hughson* (relative of Gen. Sam Houston, of Texas), b. April 16, 1842, Licking Co., O.; she was educated at Granville; moved to Fort Dodge, 1864; farmer and liveryman.

ISSUE:

1.—**Bertram E.**, b. Sept. 27, 1863; m. Josephine G. Trusty, July 20, 1885; employe of Ill. Cen. R. R.; res. Fort Dodge.
2.—**Lillian L.**, b. Oct. 22, 1865: m. May 1, 1889, A. C. Wiehe, foreman of railroad machine shops; res. Cherokee, Iowa.
3.—**Charles L.**, b. Sept. 20, 1867; m. Mollie Line, Oct. 17, 1889; liveryman; res. Fort Dodge, Iowa.

169 Henry L. Twining, son **83** Merrick, b. Sept. 16, 1841; m. *first*, Annie Moore, who d., and he m. a second wife; farmer; lived for a while in Ohio; is now res. of McDade, Texas.

ISSUE BY FIRST WIFE:

1.—**Watson**, d. young.
2.—**Lewis**, b. about 1865; m. and has a dau.; railroad conductor.
3.—**Hermon**, b. about 1868; single.
4.—**Bell**, d. young.
5.—**Mary** or Macy M., b. about 1877.
6.—**Alma**, b. about 1883.
7.—**Infant dau.**, by second wife.

170 Nelson L. Twining, son of **83** Merrick, b. Oct. 10, 1850; m. Caroline Hall, Aug. 22, 1872, b. July 19, 1854, Hartford, O ; res. Granville, O.

ISSUE:

1.—**Fred. C.**, b. May 28, 1874.
2.—**Charles M.**, b. Sept. 10, 1878.
3.—**Muriel M.**, b. July 16, 1883.

171 Edward T. Twining, son of **84** Edward W., b. Aug. 5, 1844, Logan Co., Ohio: m. Jan. 13, 1869, Florence Conger, b. Sept. 25, 1848, Washington, Iowa; physician; res. Des Moines, Iowa.

ISSUE:

1.—Leonora, b. April 13, 1870, Marshall, Iowa.
2.—Luella, b. Nov. 14, 1871, Washington, Iowa.
3.—Clark C., b. Nov. 14, 1880, Corning, Iowa.
4.—Edward H., b. Feb. 7, 1882.

172 Lauriston Twining, son of **84** Edward W., b. June 7, 1848, Iowa City; m. Laura A. Botkin, June 3, 1873, b. M'ch 12 1853, Pickway, O., dau. of Methodist minister and grad. Ill. Female Coll ; attorney and real estate; res. Des Moines, Iowa.

ISSUE:

1.—Arthur B., b. May 1, 1874, Washington, Iowa.
2.—Granville H., b. July 25, 1876, Corning.
3.—Edward L., b. Nov. 17, 1878.
4.—Einez A., b. June 21, 1882, Des Moines.

173 Jessie L. Twining, son of **84** Edward W., b. Aug. 5, 1850, Washington, Iowa; m. Flora D. Rowley (dau. Pres. minister), Oct. 25, 1876, b. Nov. 12, 1857, Peoria, Ill.; physician and druggist; res. Corning, Iowa.

ISSUE:

1.—Carrie E., b. M'ch 29, 1881, Corning.
2.—Anna J., b. Feb. 21, 1883, Corning.
3.—Jessie Lois, b. Aug. 22, 1885, Corning.

174 Hiram Twining, son of **85** Jonathan, b. June 9, 1819, Townsend, Mass.; m. Jan. 1847, Betsey

Needham, Shrewsbury, Vt. Removed to Green Garden (Will Co.), Ill., 1851, where he has lived ever since; farmer; Free Will Baptist. Commenced life poor, and after a time met with severe losses, by prairie fires, but by perseverance owns a farm of 240 acres well stocked with horses and cattle. "A small man with dark hair;" d. Dec. 14, 1889, of heart disease at his res., Frankfort, Ill.; chil. all farmers of Wills Co., Ill.

ISSUE:

1.—**Watson F.**, b. April, 1848, Shrewsbury, Vt.; res. Frankfort, Ill.
2.—**Dana**, b. M'ch, 1850, Shrewsbury, Vt.; res. Frankfort, Ill.
3.—**Leonora L.**, b. Dec. 1853.
4.—**Jasper E.**, b. Jan. 1857; m. Oct. 1884, Mary Crick. *Issue:*
   1. Henry A., b. July, 1885.
   2. Infant girl.
5.—**Irene Eliz.**, b. Feb. 1862; attending Valparaiso, Ind., Normal.

175 George F. Twining, son of **89** Jonathan, b. Jan. 28, 1838; m. Annie Whittier, of Hampton Falls, N. H., April 25, 1865; res. Charlestown, Mass.

ISSUE:

1.—**George W.**, b. Sept. 13, 1867, E. Boston, Mass.
2.—**Mabel E.**, b. Oct. 10, 1872, E. Boston, Mass.
3.—**Amy E.**, b. Dec. 31, 1874; d. July 3, 1881, Charlestown.
4.—**Edna M.**, b. April 2, 1877, Charlestown; d. Aug. 8, 1877.

TWINING AUTHORS

Elizabeth, of Troy, N. Y., a Friend, d. 1827. "Some Account of the Religious Experience and Travels," begun 1811, in nineteenth year of her age. Comly's Miscell., vol. 49, Phila., 1834.

Elizabeth, "Illustrations of Natural Orders of Plants, with Groups and Descriptions and 160 Colored Plates." London, 2 vols. Price, 1866, £21; other works.

Henry, "On Elements of Picturesque Scenery, Considered with Reference to Landscape and Painting." London, r. 8vo, vol. 1, anon, 1846, pp. 375.

Miss Louisa, "Symbols and Emblems of Early and Medevial Christian Art," London, 1852. "Types and Figures of the Bible, Ill., by Art of Early and Middle Ages," 1855.

Richard, "Pamphlets on East India Com. Topics, etc.," 1784–96.

Thomas, "Averbury in Wiltshire, the Remains of a Roman Work, Erected by Vespasian and Julius Agreote," London, 1723.

Thomas, b. 1734 and educated at Sidney Coll., Cambridge, and became rector of White Notley, Essex, 1768; rector of St. Mary's, Colchester, 1770, and d. 1804. 1. "Aristotle Treatise on Poetry and Translated with Notes," and "Two Dessertations on Poetical and Musical Imitations," Ox. 1787, with improvements by his nephew, Daniel Twining, London, 1812.

Thomas, Baptist minister, of Trowbridge, Wilts, "Sixteen Sermons with Biog.," preface by Dr. J. Toulmin, 1801, 8vo.

Thomas, "Letters on Danger of Interfering in Religious Opinions of the Natives of India," 1808, 8vo.

William, b. Nova Scotia; was educated at London; entered the medical department of the Royal Army, 1812, and d.

in Calcutta, 1835. "Clinical Ill. of the More Important Diseases of Bengal, Calcutta and London," 1832–35; two vols., 8vo.

William, M. D., d. 1848, aged 35, author, London, 1843.

Richard, Jr., " Renewal E. I. Co.'s Charter," 1813, 8vo.

Henry M., b. 1799, Penn.; prominent lawyer, teacher and writer. Author of numerous literary productions, 1850–75.

Alexander C., b. New Haven, Ct., 1801. "Astronomical and Constitutional Questions."

William, b. New Haven, Ct., 1805. "Antiphonal Psalter and Liturgies," 1877.

Nathan C., b. N. Y., 1834. " Papers on Astronomy, Language, Science, etc.," Wisconsin.

ERRATA.

Page 10: The punctuation at the word " ancestor," top of page should be omitted.

Page 23: Asterisk instead of dagger at *Annie*, foot note.

Page 40: No. 23 for 32 Thomas.

Page 42: MAHLON instead of *Malhon*.

Page 44: TAMER for *Thamer*.

Page 45: Last line, " the " should follow and not precede " considerably."

Page 70: TIBBITS for *Fibbits*.

Page 78: 41 Abner, d. Jan. 27, 1853, instead of Feb. 1, 1850.

Page 86: Hugh Twining should be numbered with "Heads of Families."

Page 102: ELLEN for *Allen*, dau. of 63 JACOB.

INDEX.

INDEX.
TO TWINING GIVEN NAMES.

Aaron, - 19, 31, 67, 85, 126
Abbott, - - 18, 19, 62, 65, 103
Abner, - - 18, 19, 51, 78, 79
Abigail, - 35, 38, 48, 49, 50, 74, 79
Abraham, - - - - 64
Abram, - - - - 143
Achsah, - - - - 157
Ada, ... 162
Addie, - - - 130, 137, 165
Addison, - - 19, 79, 123, 131
Adeline - - - - 82
Adna, - - - - - 93
Adonijah, . - - - 56
Aimee, - - - - - 157
Albert, 104, 123, 141, 145, 155, 163
Alice, ... 165, 166
Alin, ... 159
Alexander, - 18, 19, 68, 107, 112
Alfred, 18, 19, 72, 106, 111, 112, 114
- - - - 153, 162
Alice, 46, 59, 61, 81, 84, 99, 123, 136
Alma, ... 168
Almima, - - - - 119
Almira, - 68, 76, 118, 149, 160
Allen, - - - 104, 149
Alvah. - - - - 139
Alvin, - - - - 151
Amanda, - 87, 131, 146, 147, 153
Amos, 18, 19, 61, 65, 99, 105, 150, 158
Amy, - - - 100, 139, 170
Angeline, . . . 148
Antoinette, - - - 98, 153
Ann, - - - 40, 42, 61, 145
Anna, . . . 159, 169, 145
Annie, 27, 29, 47, 67, 68, 90, 93, 99
- - 116, 149, 151, 156, 157

Apphia, - - - 47
Artemus, - - - - 91
Arthur, 71, 123, 130, 132, 156, 167, 169
Augustus, - - 91, 93, 148
Austin, . 165

Barclay, ... 155
Barnabas, 14, 18, 19, 34, 38, 49, 50,
- - 74, 77, 116, 122
Bathsheba, - - - - 38
Beatrice, ... 135
Bell, . . 168
Belle, . . 162, 163
Bemsley, - - - - 121
Benjamin, 18, 19, 36, 39, 41, 57, 58
- - 59, 93
Bertha, - - - 132
Bertram, ... 168
Bessie, . . 164, 166
Betsey, - - - 59, 73, 87
Beulah, - - - 37, 62, 154
Bevel, . . 115, 164
Blanche, . . 161
Burr, . . . 137
Burton, . - 130, 164
Byron, - - - 93

Caroline, 61, 71, 101, 106, 145, 147
. 148
Carper, . . 167
Carrie, . 141, 161, 169
Cassius, . . . 164
Catherine, 75, 109, 167

INDEX.

Charles, 18, 19, 52, 54, 56, 67, 74, 82
   84, 93, 102, 107, 112, 116, 118
   126, 133, 134, 135, 136, 138, 139
   143, 145, 148, 149, 150, 155, 160
   - - - 161, 163, 164, 168
Chapin, - - - 19, 88, 130
Chapman, - - - - - 106
Chester - - - 19, 94, 140
Christean, - - 59
Christopher, - - 14, 59
Clara, - - - - 70
Clarence, . 128, 145, 158, 166
Clark, ... 169
Clarissa, - - - 140
Clifford, . . 167
Clinton, . 163, 164
Clive, . . . , 164
Cora, - - - 112, 141, 161
Cloe, - - 52, 79, 147
Cordelia, - - - 117
Corintha, - - - 70, 73
Croasdale, . 18, 19, 67, 106, 107
Cynthia, - - - 50
Cyrus, 18, 19, 59, 64, 104, 146, 158

Daisy, - - - 144
Dana, ... 170
Daniel, - - 14, 18, 41, 59
Darius, - - - 76
David, 18, 19, 33, 36, 42, 43, 50, 62
   65, 79, 83, 93, 104, 105, 139, 157
Deborah, - - .61, 67
Della, - - 138, 140
Desire, - - 141
Dewitt, - - 19, 89, 131
Dora, - - - 143
Dorcas, - - 93, 137

Earl, - - 130, 132, 156
Earmer - - - 156
Earnest, . . 130
Ebenezer, - 19, 78, 121, 122
Edgar, - - - 81, 152
Edith, . . . 99, 159
Edmund, - - 105
Edna, . 170
Edroy, 153

Edson, ... 166
Edward, 7, 19, 60, 76, 82, 109, 118
   - 119, 133, 157, 159, 167, 169
Edwin, - - 19, 106, 158
Einez, ... 169
Eleaner, - - - 136
Eleazer, 14, 18, 19, 30, 32, 37, 38, 42,
   - 48, 61, 62, 74, 100, 117, 146
Elam, - - - 150
Eli, - - 19, 99, 143, 151
Elins, - 19, 99, 145, 153
Elisha, - - 67, 93, 94
Elijah, 18, 19, 37, 38, 45, 47, 48, 73
. - - 75, 114, 117
Eliza, 27, 29, 34, 92, 108, 116, 136
   - - 140, 143
Elizabeth, 35, 37, 40, 41, 42, 44, 50
   60, 61, 63, 65, 66, 86, 99, 105, 106
   - - 122, 126, 137, 149
Ella, - - - 63
Ellen, 80, 81, 84, 88, 102, 105, 137, 156
Ellie, . . . 145, 159
Elma . . . 146
Elmer, - 143, 146, 148, 150
Elmira, - - 100, 121
Elnora, - - - 95
Elphonzo, - 19, 114, 163
Elsie, - - 138, 156
Elvira, - - 124
Elwood, - - 88, 154
Emergene, - - 112
Emily, 62, 71, 80, 82, 94, 127, 133,
   - - 140, 145, 148
Emogene. . . 163
Emoline, - - 75, 149
Emma, 142, 143, 154, 158, 165, 166
Ernest, - - - 126
Estella, - - - 132, 139
Ethel, . . . 158
Etta, - - - 140
Eugene, - - 136, 154, 164
Eva, - - - 106
Evaline, . 153

Fannie, - - 67, 106, 153
Fenimore, - - - 119
Flora, - - 141, 163

ii

INDEX.

Florence, - 80, 85, 137, 162	Ida, - - - - 137, 159
Floyd, - - - - 139	Iredell, - - - - 159
Frances, - - 65, 81, 149	Irene, ... 170
Francis, - 80, 112, 151	Isaac, 18, 19, 36, 62, 67, 100, 101
Frank, . . . 147, 164, 165	- - - - 102, 103, 156
Franklin, 19, 94, 101, 132, 136, 138	Isabel - - 23, 145, 149
- - 141, 144, 147, 155, 158	
Fred, 124, 135, 139, 140, 142 162,	
. . . . . 166, 168	Jacob, 18, 37, 42, 44, 60, 62, 64, 95
Frederick. - - 19, 59, 94, 164	97, 98, 102, 104, 146, 147, 148, 149
	James, 19, 59, 60, 78, 85, 95, 97,
	- 105, 115, 127, 135, 145, 148
Geneviere, ... 163	Jane, - 81, 85, 95, 145, 156, 160
George, 18, 19, 65, 94, 98, 105, 112	Jasper, ... 170
125, 127, 140, 141, 151, 152, 156	Jay, - - - - 151
. . . 160, 162, 166, 170	Jennie, - - - 141, 151, 158
Georgiana, - - - - 152	Jeannetta, - - - - 119
Gertrude, - - - 139, 143	Jeremiah - - - - 94
Gleene, ... 162	Jessie, 18, 19, 64, 94, 104, 114, 119
Glendora, - - - 138	- - - - 143, 169
Grace, ... 162	Joanna, - - - - 29
Graff, - - - - 148	Joel, - - - - 78
Granville, . . . 169	John, 9, 18, 19, 30, 31, 32, 36, 38
Gratia, - - - - 119	40, 41, 44, 50, 55, 56, 57, 58, 59
	60, 62, 81, 83, 84, 85, 87, 88, 91
Hallowell, - 19, 99, 101, 153, 155	92, 95, 96, 97, 103, 105, 106, 113
Hannah, 34, 38, 42. 43, 50. 51, 59	131, 136, 141, 143, 145, 146, 147
95, 97, 106, 124, 125, 148	149, 150, 152, 156, 157, 158, 161
Harriett. 108, 114, 118, 149, 153	- - - - - 164
Harrison, - - 19, 79, 124	Jonathan, 18, 19, 38, 49, 51, 52
Harry, - 128, 144, 148, 152, 166	- 77, 79, 98, 105, 119, 151, 158
Hattie, ... 165	Josanna, - - - - 144
Helen, - - 68, 111, 112	Joshua, - - - - 99
Henrietta, - - - 118	Joseph, 18, 33, 36, 40, 41, 49, 58
Henry, 18, 19, 43, 58, 61, 64, 73	60, 73, 75, 81, 93, 94, 96, 97, 114
87, 94, 104, 116, 118, 128, 142	124, 125, 137, 138, 140, 145, 146
- 144, 145, 146, 149, 168	- - 155, 157, 163, 164
Herbert, - - - 124, 139	Josephine - - - 81
Hermon, ... 168	Josie, - - - - 139
Hester - - - 41	Judah, - 18, 19, 49, 74, 75
Hiram, 18, 19, 73, 115, 116, 143, 169	Judith, - - 93
Homer, - 19, 45, 117, 150, 167	Julia, - - - 72, 116
Horace, - 19, 95, 101, 144, 155	
Howard, - 19, 104, 124, 154, 157	
. . . . 159, 167	Katie, - - 142, 143, 164
Howe, - - - - 137	Kinsley, - - 14, 19, 108, 159
Hugh, - - - - 85	Kossuth, - - - - 141
Huldah, - - - 143	

iii

INDEX.

Lambert, 147	Martha, 50, 58, 59, 60, 97, 101, 102
Laura, 161	103, 107
Lauriston, - 19, 76, 119, 169	Marvin, - - 85
LaVerne, - - 140	Margery, - - 106
Leah, - - - - 92	Mary, 7, 35, 37, 39, 40, 41, 42, 44, 56
Leland, - - - - 136	59, 60, 61, 64, 67, 68, 73, 78, 81
Lena, - - - 135, 166	82, 88, 89, 91, 92, 93, 95, 96, 98
Leonard, - - - - 132	99, 100, 101, 104, 105, 106, 107
Leon, - - 126, 127	108, 111, 115, 116, 122, 136, 137
Leonora, - 132, 169, 170	141, 145, 146, 148, 151, 155
Leota, - - - 167	162, 168
Letitia, - - - - 61	Maryette, - - 91
Lewis, 18, 19, 49, 75, 76, 90, 113	May, - - 157
115, 117, 118, 133, 162, 166	Maud, - 139, 142, 143, 150
167, 168	Melissa, - - 90, 131
Lillie, - - 130, 159, 160	Medora, - - - 116
Lillian, - - - 114, 168	Mercy, - 30, 32, 34, 40, 44, 50, 98
Lionall, - - - - 130	Meribah, - - 122
Lizette, - - - 124	Merinda, - - 121
Lizzie, - - - - 157	Merrick, - 19, 76, 118
Lois, 49, 75, 118, 164, 166, 167	Mertie, - - 136
Loise, - - - - 47	Millie, - - 146
Louisa, - - 71, 102, 106	Milo, - - 74, 114
Lucinda, - - 73, 113, 150	Minnie, - 106, 139, 141, 167
Lucius, - - - 19, 115, 165	Mollie, - - - 160
Lucina - - - 121	Mont, - - 156
Lucine, - - - 167	Moses, - - 131
Lucy, - - 52, 81, 124, 142	Muriel, - - 168
Luella, - - - 169	Musetta, - - 156
Lula, - - - 135, 144, 151	
Lycurgus, - - - 73	Nancy, - - 91, 93
Lydia, - 49, 51, 70, 97, 142, 143	Nathan, 18, 19, 49, 77, 79, 86, 123
- - - - 149, 154	124, 128, 129
Lyman, - - - 73, 80	Nathaniel, 14, 18, 30, 31, 32, 35, 36, 40
	Nellie, - 137, 139, 143, 165
	Nelson, 19, 74, 117, 118, 162, 168
Mabel, - - 170	Nettie, - - 67, 167
Maggie, - - 154	
Mahlon, 18, 19, 42, 43, 58, 62, 94, 142	Olive, - 91, 124, 139
Mahitable, - - 27, 29	Oliver, - - 58
Malichi, - - 19, 96, 144	Orlandon, - - 19, 114, 164
Malvina, - - 122	Orlow, - 163
Mamie - - - 152	Ovanda, - 161
Margaret, - 35, 85, 86, 146	
Maria, - 105, 112, 145	
Mariamna, - - 93	Paschal, - - 121
Marietta, - - 114	Paul, - 105, 131

iv

INDEX.

Paulina,	73
Pearl,	166
Perry,	131, 139
Peter,	19, 87, 130
Phebe,	62, 96, 100, 140, 147, 149
Philander,	19, 75, 117
Philino,	74, 115
Phineus,	86
Philip,	19, 92, 137
Polly,	50, 74
Preston,	163
Prince,	18, 38, 45, 52, 80
Priscilla,	64
Prudence,	92

Rachel,	32, 37, 41, 43, 44, 55, 58, 61, 70, 90, 92, 93, 103, 104, 106, 132, 156
Ralph,	19, 97, 130, 144, 149, 150, 160
Ray,	130
Rebecca,	58, 71, 82, 92, 99, 100, 121, 154
Reuben,	100
Richard,	14
Riley,	100, 142
Robert,	101, 142, 146, 146
Romulus,	73
Rosamond,	139
Roscoe,	161
Rose,	130
Rozetta,	93, 138, 132
Rozilla,	51
Russell,	145, 153
Ruth,	34, 38, 49, 61, 74, 104, 117
Ruthanna,	100

Sabra,	51
Sabrina,	122
Sallie,	105, 106, 152, 157
Samuel,	14, 18, 19, 36, 39, 40, 55, 58, 61, 70, 92, 93, 95, 100, 115, 116, 144, 150, 156, 164, 166
Sarah,	37, 40, 42, 53, 58, 60, 63, 64, 73, 79, 82, 84, 85, 95, 96, 102, 103, 104, 107, 108, 121, 124, 126, 132, 137, 138, 139, 147, 149, 156, 163

Selinda,	54, 56, 90
Seth,	80
Seymour,	137
Silas,	18, 19, 42, 61, 97, 100, 150, 153, 154
Sophronia,	72
Stephen,	18, 19, 28, 29, 30, 31, 32, 35, 37, 38, 39, 44, 45, 53, 65, 67, 72, 82, 106, 126
Sukey,	125
Susan,	84, 85, 151, 156, 157
Susanna,	29, 43, 49, 64, 98, 103, 113
Sutherland,	108
Sybil,	164
Sybilla,	135

Tabitha,	50, 77
Tamer,	44
Tamzin,	79, 125
Ted,	156
Thankful,	34, 51, 79, 94
Theodore,	108
Thomas,	9, 14, 18, 19, 37, 38, 40, 45, 48, 49, 50, 53, 54, 55, 56, 57, 58, 62, 64, 70, 84, 88, 89, 91, 92, 101, 103, 104, 105, 132, 135, 136, 146, 157
Thompson,	153
Timothy,	38, 49

Uriah,	19, 100, 154

Violet,	123
Virgil,	139

Walker,	157
Walmsley,	105
Walter,	135, 154, 157
Warren,	85
Watson,	18, 19, 61, 99, 145, 153, 168, 170
Wesley,	146
Wilber,	151
Wilkinson,	145

INDEX.

William, 10, 14, 18, 19, 20, 21, 22, 23, 25, 26, 27, 28, 29, 33, 34, 35, 37, 38, 46, 48, 57, 59, 60, 62, 68, 69, 73, 82, 91, 92, 94, 95, 100, 102, 104, 105, 106, 107, 108, 109, 110, 111, 112, 113, 114, 124, 127, 135, 136, 139, 140, 141, 144, 145, 146, ... 147, 148

William, 149, 150, 152, 153, 154, 156, 158, 159, 160, 161, 162, 163, 164
Williamina, . . . 147
Williamson, — — 79
Williby, — — — 146
Wilmer, — - 105, 158
Winthrop, . . . 163

OTHER NAMES THAN TWINING.

Adams, . 97, 113, 137, 149	Bennett, . 40, 57
Alerson, . . . 62	Bent, . . 162
Ames, . 93, 137	Betts, . 127
Amond, . 149	Bills, . . 27, 29
Amos, . 62, 100	Birdsell, 94, 140
Andrews. . 165	Blaker, 103, 148
Apger, 151	Boise, . 117
Ashley, . 119	Bosler, . 146
Ashton, 101, 155	Botkin, . 119, 169
Atkins, . 58, 93	Boutwell, . 56, 87
Atkinson, 7, 64, 65	Bradfield, . . . 40
Atwood, . 52	Briggs, . . 41, 103, 157
	Brooks, . . . 145
Bacon, . . 107	Brown, 46, 69, 71, 81, 115
Baker, 23, 30, 142, 162	Buckley, . . . 139
Balch, . . . 92, 135	Bucklin, . 112
Balderston, . 41	Buckman, 42, 64
Baldwin, 39, 53, 137	Bunson, . . 39
Balliel, . . 98, 151	Bunter, . 39
Barber, . 127, 128	Bunts, . 140
Barclay, . 27	Bunting, - 30, 83
Barton, . 106, 111	Burr, . . 146
Barrick, . 119	Burroughs, . 98
Bartleson, . 101	Byers, . . 112, 160
Bascom, 47, 77	
Bates, 138	Calkins, . 140
Battey, . . 90	Callison, . . 132
Baynes, . 101, 155	Campbell, 101, 111, 160
Beach, . . . 72	Carman, . 94, 141
Bean, 61, 101	Carpenter, 112, 160
Beans, . 64, 104	Castle, . . 91
Bechtold, . 105	Catlin, . 46, 67
Beecher, 70, 139	Chapman, 33
Benedict, . 135	

vi.

INDEX.

Chickering.	162
Choat,	92
Clark,	76, 117, 118, 121
Clayton,	47, 77
Cleveland,	56, 57, 58
Cobb,	29, 33, 50, 77, 122, 123, 124
Colson,	79, 123
Colton,	165
Cook,	124
Cone,	114, 163
Conger,	119, 169
Cooledge,	125
Cooke,	25, 26, 32
Coon,	91
Cooper,	67, 87, 105, 130, 157, 163
Cordill,	132
Cotanche,	91
Councilman,	92, 137
Cowan,	166
Coy,	163
Cramer,	162
Crane,	73
Crick,	170
Croasdale,	44, 67
Crocker,	135
Cromwell,	10
Crook,	40
Crosby,	78
Cundiff,	101
Dakin,	80
Davis,	30, 81, 92, 146
Daws,	36, 41
Deane,	23, 25, 26
Decoursey,	145
Defan,	145
Dimon,	92
Doane,	23, 29, 45
Dodge,	4, 72
Downs,	79
Duffil,	60
Eastburn,	82
Eberman,	65, 105
Ebert,	146
Eckert,	105, 158
Eddy,	80
Eleman,	105
Eldridge,	121
Elliott,	35
Elmer,	164
Ely,	98
Ervin.	164
Evans,	28, 93
Fargo,	114, 162
Farthing,	146
Fessenden,	77
Field,	40
Fields,	85
Filtenberger,	149
Filley,	69, 70
Firman,	148
Fish,	143
Fluke,	166
Fonner,	58, 92
Fosdick,	166
Fotton,	140
Fox,	26, 80, 139
Fowler,	47, 48, 49, 73, 75, 113
Frankenfield,	103, 156
Franklin,	22, 25, 27, 143
Frasier,	149
Funston,	64
Galbreath,	131
Galord,	142
Gano,	60
Gates,	115, 163, 164
Gaylor,	81, 94
Getty,	62
Gibson,	138
Gibbs,	92, 136, 162
Gifford.	95, 143
Gilkey,	124
Gilland,	43
Girdler,	142
Goss,	139
Goodwin,	143
Goodseede,	58, 94
Goodsell,	49
Gould,	50
Graham,	46, 123
Grant,	55
Graves,	146

INDEX.

Gridley,		108	Hudson,	121
Griffith,		85	Hughson,	118, 168
Grimore,		124	Humphreys,	112
Grindle.		123	Hunt,	95
			Hutchinson,	21, 37
Hadley,		68		
Hall,	49, 118, 168		Ingelow,	4
Hallowell,	61, 62, 99, 100		Ingham,	8
Halliwell,		146	Iredell,	106, 158
Hambleton,	88, 89, 90, 131		Irwin,	112
Hamilton,	35, 72, 111		Irvin,	114
Hamlin,		167		
Hampton,	83, 85, 145		Jackson,	115, 165
Harding,	37, 42, 62		Jacoby,	102
Harling,		8	Jansey,	35
Harris,	41, 95, 144		Jaquett,	63
Harrison,	72, 74		Jenks,	36, 39, 40, 145
Harrold,	61, 99		Jerome,	117
Harper,		94	Johnson,	56, 68, 98, 109, 165
Hart,		98	Jones,	43, 58, 70, 95, 118
Hartman,		59	Jonson,	93
Harvey,		148	Josephus,	16
Hass,		59		
Hawn,		161	Kee,	88, 130
Hayes,		141	Keyes,	94
Heaston,	97, 150		Kendall,	102
Heath,		137	Kendrick,	50
Hedge,		24	Kennedy,	60, 148
Hellyer,		97	Kenyan,	115
Hendricks,		51	Kerboah,	39
Henry,		73	Kester,	56, 88, 89, 131
Herbert,		139	Kickok,	58
Hesser,		137	Kilfoil,	167
Hicks,		36, 40	Kinsley,	68, 107
Higgins,	38, 45, 49, 50, 51, 52, 79, 80, 122		Kirk,	32, 35, 36, 67, 99, 103, 106, 157
Hill,		124	Kirkbride,	28
Hillborn,		35	Kite,	106
Hoagland,		104	Klette,	39
Holland,		125	Knapp,	95
Homer,		150	Knight,	79, 153
Hopkins,	37, 50		Knowles,	45, 50
Hoskins,		164	Krusen,	65
Hotten,		10	Kyle,	104, 158
Houston,		168		
Houghton,		115	Lacy,	61, 100
Howard,	92, 135		Laing,	55
Howe,		137	Laird,	63

INDEX.

Lambert,	147	More,	73, 114
Lance,	59, 94	Morehouse,	164
Laning,	59	Morgan,	88, 145, 166
Large,	82	Moody,	109
Lashier,	58	Moore,	74, 118, 167, 168
Laud,	10	Moses,	147
Leavenworth,	159	Naylor,	32
Lebenan,	92	Needham,	21, 170
Lee,	41, 60	Newcomb,	112
Leedom,	37	Nickerson,	38, 47, 50
Lewis,	34, 36, 37, 73, 84, 116, 153, 162	Nippes,	101, 155
Lindsey,	154	Norris,	132
Line,	168	Norrigan,	147
Lininger,	132		
Linnell,	79, 124	Oglebee,	106
Lisk,	95	Omsted,	107
Liverzey,	146	Overstreet,	116, 166
Lomison,	59	Owen,	81
Loosely,	96		
Lupton,	32	Page,	154
Luther,	91	Paine,	7
Lynch,	136	Palmer,	44, 55, 83
Lyons,	85, 126	Parker,	54
		Parsons,	75
Mahan,	147	Patten,	161
Magill,	44, 68	Patterson,	99
Manchester,	90	Payne,	94, 141
Marsh,	122	Peace,	73, 115
Maranville,	116	Penn,	27
Marshall,	109, 118	Penquite,	32, 35
Mathews,	147	Perry,	86
Mayo,	22, 39, 45, 51	Phelps,	74, 116
McBride,	146	Philo,	95
McCullick,	94, 142	Pickett,	71, 75
McDowell,	42	Pidcock,	103
McDonald,	70, 107	Pierson,	118
McKinzie,	58, 91	Platt,	25
Meachem,	81	Plumket,	108, 159
Meeker,	140	Potter,	81
Megadigan,	103	Powers,	57
Merrell,	163	Prence,	22, 24
Mettler,	59	Price,	71
Miller,	37, 42, 87, 113, 128, 166		
Milton,	10	Quick,	166
Mills,	113		
Miner,	92, 136	Ramsey,	147, 164
Mitchell,	32, 35, 114, 127	Ramson,	147
Molloy,	145	Ranchler,	149

INDEX.

Randolph,	101	Sisco,	95, 144
Raidick,	165	Slack,	105, 158
Rayne,	141	Slocum,	72
Reed,	135	Slutter,	152
Reeder,	98, 147	Small,	78, 121
Rennie,	127	Smith,	48, 49, 50, 64, 73, 74, 75, 113,
Reynolds,	135		114, 116, 121, 133, 148, 161
Ribble,	59	Snow,	50, 51, 78, 79
Ridge,	154	Snyder,	41, 58, 149
Riley,	62, 100	Spaulding,	121
Ring,	26	Spoffer,	84
Rittenhouse,	128	Spencer,	164
Roach,	85	Sperry,	109, 159
Roberts,	92, 136, 147	Sprague,	90
Rockwood,	114	Staley,	97, 148
Robeson,	55	Stanton,	91, 133
Robinson,	76	Stackhouse,	64
Rogers,	28, 29, 37, 38, 47, 49, 50	Starner,	91
	51, 77, 114, 116	Steese,	51
Rood,	81	Stephens,	106
Rook,	161	Stevenson,	142
Rose,	76	Stewart,	115
Rounds	93	Stoddard,	81
Rowland,	30, 81	Stoffer,	84
Rowley,	119, 169	Stokes,	44
Rush,	95, 144	Stout,	56, 57, 58, 93
Ryan,	62, 102	Strickland,	74
		Sweet,	34, 38
Samuels,	49	Switzer,	95, 143
Samvels,	49		
Savage,	21, 25	Talcott,	75, 115, 163
Savidge,	102, 155	Tacitus,	12
Scaife,	28	Taylor,	8, 43, 65, 94, 112, 114, 143, 161
Scarborough,	67, 98	Tenny,	108
Schele de Vere,	8	Terry,	135
Schermerhorn	93, 138	Thatcher,	95
Schofield,	82	Thomas,	86, 122
Schwartz,	131	Thompson,	113, 151
Scott,	60	Thomson,	153
Seaver,	84	Thorn,	109, 132
Seeley,	90	Tibbits,	70
Sharps,	106	Tinker,	73, 112
Shepard,	73, 75, 112, 117	Tomlinson,	41, 60, 61, 65, 105
Sherman	90, 133	Torbert	37
Sherwood,	88	Townsend,	81, 95
Shoemaker,	99	Treat,	25
Simons,	145	Trego,	67
Simpson,	53	Tripp,	94

x

INDEX.

Trusty,	. . . 168	Whelpley,		113
Tucker,	42, 60, 62, 95, 97, 98, 100	Whipple,		51
Tuttle,	. . 122	Whit, .		103, 156
Twyn,	8	Whitaker,		. . 130
Tyson,	99, 153	White,		7, 72, 117, 167
		Whitman,		. . 95
Vanartsdalen,	100, 154	Whittier,		125, 170
Vandegrift,	. . 105, 158	Wiehe,		168
Van Horn,	. 40, 65, 95, 100, 154	Widdifield,		55, 87
Venumn,	. . 162	Wilson,		30
Vicking,	8	Wildman.		. 39
		Wilkinson,		37, 39, 44, 67
Waldo,	116, 162	Williams, .		99, 118, 146, 153
Walker,	82, 88, 126	Winner,		. 40
Walter,	. 153	Wise,		59
Wardwell,	. . . 58, 79	Wismer.		150
Warner, 62, 67, 82, 103, 105, 106, 126		Woods, .		121
. . . . 148, 158		Woodman,		. 79
Ward,	. 30	Woolcott,		47, 49, 167
Waring,	8	Woolsey, .		107
Waters,	. 73	Worth, .		. 39
Watson,	. 44, 102	Worthington,		. 103, 145
Webb,	84, 117, 166	Wright,		. 100
Webster,	. 56, 88			
Weed, .	76, 85, 119			
West,	. 53, 82	Young,		25, 29, 30
Welding,	42, 61			
Wheldin,	163	Zell,		15

The Compiler will deem it a favor should the reader freely suggest errors that may occur in the book; and he will be pleased to receive whatever additional information may be offered, or new facts obtained, regarding the families, name or connections thereto, treated in the preceding pages.

www.ingramcontent.com/pod-product-compliance
Lightning Source LLC
Chambersburg PA
CBHW032151160426
43197CB00008B/857